Waterwise House & Garden

A Guide for Sustainable Living

Allan Windust

LAND
LINKS

National Library of Australia Cataloguing-in-Publication
Windust, Allan.
Waterwise house and garden: a guide for sustainable living.
ISBN 0 643 06800 7
1. Landscape gardening – Water conservation – Australia.
2. Water conservation – Australia – Citizen participation.
I. Title.

333.91220994

Copyright © Allan Windust 2003

Published by:
Landlinks Press
PO Box 1139
Collingwood Vic. 3066
Australia

Disclaimer
While the author, publisher and others responsible for this publication have taken all appropriate care to ensure the accuracy of its contents, no liability is accepted for any loss or damage arising from or incurred as a result of any reliance on the information provided in this publication.

Foreword

There is no doubt that Australia is a dry country where water plays an important role in growing plants, particularly at present. We use a lot of water in our homes, especially on our gardens. Most parts of Australia have dry periods that can extend to become droughts. Right now many cities are experiencing problems maintaining their water supplies. These problems will continue into the future unless we act now. We can all use less water in our houses and gardens.

Allan Windust shows many ways we can reduce our water consumption and still live well. He offers a range of options – some require little effort beyond changing our everyday habits in small ways that, in the end, add up to significant water savings.

I like the way he explains the science behind gardening and how plants use water. This gives us insights into how we can become waterwise gardeners.

The book mentions many plants that you could find useful in your area, but remember to talk with your local nursery about such plants.

Allan has given a lot of thought to the management of a waterwise house and garden. Well done Allan – I'm sure readers will be rewarded with your many hints. Always remember: gardening must be relaxing, not something to worry about.

Happy gardening!

Kevin Heinze AM

Kevin and Allan with the eucalypt that Kevin planted in his garden over 40 years ago.

Contents

Acknowledgments   ix

Chapter 1   Introduction   1

Chapter 2   The value of water   3

Chapter 3   The importance of plants   5
    Work with nature   5
    A positive experience   6

Chapter 4   The house and garden system   7
    Aim to survive and go on surviving   7
    Spin-offs   8
       The suburban carbon sink   8
       The suburban aviary   9
       Salinity   9
       Dollars and cents   11
    Australian climate   11
    Australian droughts   11

Chapter 5   Waterwise strategies   15
    Choose your approach   15
    Know your climate   15
    View the house and garden as a system   16
       Rainfall   16
       Mains water   17
       Garden evapotranspiration and seepage losses   17
       Household losses   18
    Become water-conscious then a water-saver   18

Chapter 6   Waterwise options   19
    Household reticulation options   19
       Toilet   19
       Shower   19
       Washing machine   20
       Laundry sink   20
       Hand basin   20
       Kitchen sink and dishwasher   20
       Bath   20
    Garden watering methods and appliances   21
       Bucket using mains water   21
       Hose using mains or tank water   21
       Sprinklers using filtered mains water   22
       Misters using filtered mains water   24
       Porous (sweat) hoses using filtered low-pressure mains or tank supply   25
       Dripper hoses using filtered mains or tank water   26
       Individual drippers using filtered mains or tank water   27
       Filters for mains supply or tank water   27

Timers   28
Moisture meters   28
Automated systems using mains pressure systems   29
Pot plants   29
On-site catchment and storage in water tanks   30
   Regulations   30
   Uses   31
   Capacity   31
   Drinking water   32
   Installation   32
   Tank construction   33
   Inspection and maintenance   33
   Elevated water tank   34
   On-ground water tank   34
   Your swimming pool   35
   Underground water tank   35
   Tank water for toilet cisterns   36
   Ponds   37
Recycling household wastewater   37
   Recycling untreated grey water onto the garden   39
   Washing machine and washing tub grey water   40
   Showers: clear or grey water   40
   Baths: grey water   40
   Hand-basins: grey water   41
   Kitchen sink or dishwasher: grey water   41
   Septic tank effluent: black water   41
   Treatment and reuse of grey water   41
   Use of treated grey water on gardens   48
   Grey water reuse conclusions   48
Garden watering methods   49
   Moisture testing   49
   Frost   50
   Heat stress   50
   Foliage or root watering   51
   Lawns   51
Water budget   52
   Watering times   52

Chapter 7   The theory and practice of mulching   53

Insulation   54
Mulch as habitat and food   55
Worms in soil without mulch   56
Worms in soil with mulch   56
Soil protection   56
Weed suppression   57
   Weeds as green manure   57
Mulch as cosmetic ground cover   57
Applying mulch   58
   Vegetables and flowers   59
   Trees and shrubs   62
   Non-organic mulches   62

      Living mulches   63
      Self-mulching plants   64
  Mulch materials   65
      Garden and household collectables   65
      Outside collectables or purchases   66
      Mulch as habitat for pests   67
      Mulch materials list   68

Chapter 8   Planning your waterwise garden   83

  Your aims   83
  Resources   84
      Finances   84
      Time   84
      Attitude   85
      Labour and equipment   85
  Site assessment   85
      Site plan   85
      Your garden climate   87
      Topography: drainage   91
      Existing plants   92
      Impediments and assets   92
      Site assessment plan   93
  Design   95
      Activity areas and linkages   95
      Dry zones and wet zones   95
      Fire buffer zones   97
      Examples of wet/dry zone fire-retardant designs   97
      Model kitchen garden   99
      Garden water recycling   101
  The master plan   102

Chapter 9   Plants   105

  Your prize plants   105
  Survival plants   105
      Australian plants   106
      Plants of the world   107

Chapter 10 Help with plant selection   109

  Your local water authority   110
  Your local indigenous nursery   110
  Your local botanic garden   110
  Public parks and specialist gardens   111
  Societies for growing Australian plants   111
  On-line   112

Chapter 11 What to do during a drought   113

  Essentials   113
  Develop a routine   114
  Start the waterwise design process   114
  Select construction projects   114
  Determine not to forget   114
  Look after the animals   115

Things to try   115
    Using pots to advantage   116
Golden rules for the waterwise gardener   116

Chapter 12 The future   119
Black water perspective   119
Grey water   120
    Attitudes   120
    Public health   121
    Damage to the environment   122
    Technology   122
Stormwater   124
Desalinisation   124

Appendix 1 The importance of water to plants   125
The whole plant   125
    The healthy plant   125
    The stressed plant   125
Soil   126
Roots   127
Stem   128
Leaves   128
Flowers and fruit   129
Plant cells   129
Plant strategies to overcome dry conditions   130
    Roots   130
    Stem   131
    Leaves   132
    Overall plant strategies   133

Appendix 2 Australian plants tolerating very dry conditions   135

Appendix 3 Exotic drought-tolerant plants   145

Appendix 4 Fire-retardant species   161

Appendix 5 Drain stranglers and cloggers   163

Appendix 6 Wastewater reuse EPA guidelines   165

Appendix 7 Water audit   167
Indoor usage   167
Outdoor usage   172

Appendix 8 Publications   181
Australian natives   181
Waterwise garden books   181
Herbs   182
Earthworms   182
Water reuse   182
General   183

About the author   185

Acknowledgements

A book such as this could not be written without the help of experts in the field of efficient water use.

My thanks to the following specialists with the various water authorities for giving me their time, access to information and comments on the text of this book:

- Bruce Rhodes and the staff at Melbourne Water
- Des Horton and his associates at City West Water, Elio Comello of the Habitat Trust and the staff of Basaltica demonstration garden
- Keith Johnson at South East Water
- Barry Jepperson of Brisbane City Council.

I am grateful to the Urban Water Research Association of Australia for allowing the publication of information and diagrams on domestic grey water reuse.

Thanks also to Polymaster water-tank manufacturers for information and photographs of water tanks.

My gratitude goes as well to staff of Australia's various botanic gardens for help with waterwise plant information. In particular, the staff of the Royal Melbourne Botanic Garden for allowing me to photograph the dry and arid garden sections.

I also wish to express my appreciation to Dr Simon Toze and Bradley Patterson at CSIRO Land and Water for access to their research into water reuse in South Australia and Western Australia. Finally, I would like to thank Ted Hamilton of CSIRO Publishing for his help and advice in seeing this work through to completion.

The help of all these experts made my path easier.

Allan Windust

Chapter 1

Introduction

Waterwise sense is a combination of common sense and common science

This book will show you how to cut your water bills by half or more, and have the garden you want with the water usage you can afford. This is based on a better understanding of water economies, soils, plants, climate and the resources available on your property and elsewhere.

This book covers all aspects of dry-weather gardening and household water usage, and emphasises understanding the forces of nature, sensible design, plant selection and management.

Since I wrote *Drought Garden* in 1994 the influence of climate change on our weather patterns has become more apparent. The action of excess carbon dioxide in our atmosphere is now recognised by most scientists involved in the study of the world's weather as the cause of climate change. There is more carbon dioxide in the air than there has been for 400 000 years. The greenhouse effect will remain with us for the foreseeable future and we face the prospect of it getting worse.

The general scientific consensus is that our planet will face increasing extremes of flood and drought. Wet places will get wetter and dry places will get drier.

Australia, with its already low rainfall, will be badly affected. We must prepare for the reality of dwindling water supplies and the impact this will have on our primary production as well as our homes and gardens.

The map and table in Figure 1.1 show a range of scenarios devised by CSIRO scientists to help formulate policy to address climate change. It shows the likely variations in rainfall across Australia as a result of global warming. The map and table are part of a large independent scientific report – *Australia State of the Environment* – which was presented to the Minister for the Environment by the State of the Environment Advisory Council.

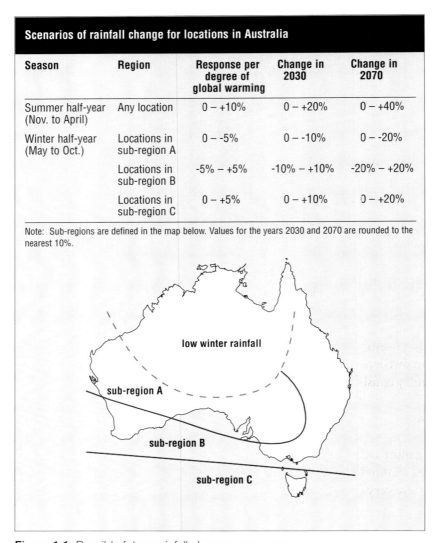

Figure 1.1: Possible future rainfall change. Source: CSIRO

If the prospect of climate change is not enough to influence garden practices, motivation will certainly come with the rising cost of water.

Waterwise House and Garden shows you the latest in water-saving technology, ideas for plantings and sustainable garden design.

Chapter 2

The value of water

Ask any Australian farmer what their most critical resource is – the unhesitating answer will be 'Water'. At any moment somewhere in Australia a landholder will be facing the challenge of a water shortage. Ours is a land of cyclic droughts that move around the country until everyone experiences one.

Urban dwellers spend less time watching the horizon and weather forecasts than their country cousins. They don't have to collect drinking water off their roofs into tanks of limited capacity. Water comes in a pipe and they can have it any time they need it. But while urban people are not directly dependent on the land and the crops and livestock that grow upon it, their existence still depends on water. Water shortages are becoming more frequent as suburbs expand; climate change adds to the uncertainty.

We are all moving toward the same level of consciousness and need to be waterwise.

There are very strong reasons for us to be water-conscious and they all relate to costs: direct costs to ourselves, costs to farmers and costs to Australia's environment.

Cities are already overusing the country catchments for water storages. The cost of constructing large dams and the damage to the environment make it highly unlikely that such dams will be built in the future to meet the needs of our ever-expanding cities. Other methods of supply, such as desalination of seawater, are prohibitively costly at the moment.

Farmers who rely on irrigation supplies are paying increasingly more for water and at times are denied water altogether. As irrigation water becomes more expensive, so does the food that farmers grow for the rest of the community.

The river environments that survived naturally for millions of years are now suffering all kinds of degradation. Algal blooms are killing our waterways, fish and other water creatures and plants are dying, weeds are clogging formerly free-flowing streams that have been reduced to gutters of trickling water.

When we overuse water we rob not only the countryside – we rob ourselves and our environment as well.

Too much water

Conversely, our water authorities have a major problem due to too much water. Too much water is being used by our waste disposal systems. Authorities are having great difficulty dealing with the mounting effluent flows from our urban areas. A major challenge of this century will be to reduce the volume of effluent entering the wider environment and to treat and reuse it so that it becomes a resource instead. Ordinary households will have a part to play in solving this problem by reducing the amount of water leaving the property as waste.

Being water-conscious

With a minimum of effort you can halve your home water usage almost immediately by adopting the following principles.

- Cut out wasteful water use practices in the house and garden.
- Understand how water acts in soil.
- Learn how plants use water.
- Learn which plants use less water and which use more, and manage them accordingly.
- Understand and apply mulch correctly.
- Know and monitor your water resources.
- Water towards dusk and during the night.
- Do not overwater.
- Recycle household grey water onto the garden.

Saving water

With more dedication, you can reduce your watering by up to a further half the remaining usage in a relatively short time. Although adopting some of the technology initially costs money, you will save on your water bills.

- Apply waterwise garden design principles.
- Know and adapt your garden management to your climate.
- Reduce lawn areas to your essential needs.
- Make maximum use of paving or gravel and make sure that rainfall runoff drains onto lawn and garden areas.
- Know your entire actual and potential water resources.
- Use waterwise technology.
- Gradually replace thirsty plant species with dry-tolerant plants.
- Involve the whole household.

Chapter 3

The importance of plants

We all know that plants look good and provide shade, but many of us seem unaware that without them humanity would never have come into existence and without them we would cease to exist.

When plant life emerged 630 million years ago the earth was uninhabited by animals. Why? Because there was no oxygen.

The primary role of plants is to extract the carbon from carbon dioxide for their growth and expel oxygen as a waste product. Yes, we live on the waste product of plants. But the miracle of plants does not end there. We also eat the leaves, roots and fruits of plants, and many of the animals that live off them. So we are absolutely dependent on plants.

This book will help you become a creator of beauty and sustainability.

Work with nature

Learning from and working with nature is the golden rule of gardening. And in dry times that rule is particularly true. What we need is calm, careful consideration.

- We cannot control the weather but we can control the climate near our plants – the micro-climate.
- We can choose the right plants for dry times.
- We can protect most plants from extremes with a little extra care.
- We can have the garden display we would like this year, and every year.
- We may have to sacrifice some plants or even sections of the garden while we concentrate on growing the plants that can survive dry conditions.

If your garden is small enough or if you have the time and resources you can achieve these goals with relatively little effort.

A positive experience

Living with less water, even surviving a drought, can have positive aspects. The main benefit is learning from observation and experience. Early in my gardening career, in western Victoria, there was a severe drought. I mulched what vegetables I could – mainly tomatoes – and left some of the others to their fate. Because I fertilised and mulched those few tomatoes well and watered by bucket, we had the best crop ever. The yield was up, the quality was great and the tomatoes ripened earlier and fruited longer. I learned a lot about mulch and water economy.

Chapter 4

The house and garden system

As we shall see, a waterwise person comes to view the house and garden as one system. Each relies on the other. The house provides a source of recycled or catchment water for the garden. The garden provides the house with a modified climate and a place of pleasure to look out on or wander in. The garden can also be a source of nourishment for the householders.

The garden climate, or micro-climate as we call it, is part of the larger climate of the district influenced by structures such as the house, sheds and fences.

The household water supply system is part of an overall supply system. It also contains subsystems within the property such as the sewerage, the drinking water subsystem and the garden system. Part of the waterwise householder's strategy is to make these subsystems work better together.

Water recycling is a major means of integrating the subsystems to save water. Capturing and storing water that falls on the roofs of the house and other buildings has important potential (see also p. 16).

Aim to survive and go on surviving

Your immediate waterwise aim is to have a garden that thrives and pleases you through good and dry seasons. Your powers, your judgement and your patience may be tested, but by using the suggestions in this book you can help your garden survive.

You must use all the resources at hand. You must continuously monitor your plants, the level of moisture in the soil, your water supply and the weather forecast. You must choose what is to survive and what to let die.

Your long-term aim is to be prepared for a future with less water, by utilising good design and good technology, and by growing the appropriate plants.

8 | Waterwise House & Garden

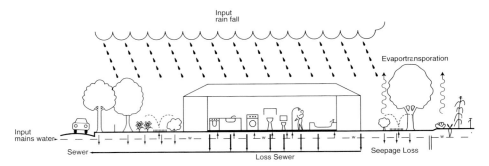

Figure 4.1a: The conventional house and garden system. Note the inputs and losses, especially seepage.

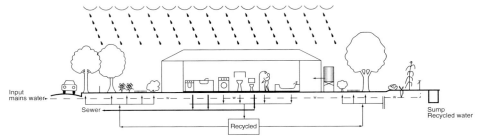

Figure 4.1b: The ideal waterwise house and garden system. Note the saving of roof runoff and reduction in soil seepage loss.

Spin-offs

A spin-off from the following activities will be a garden that will help sustain the local, national and indeed the world environment.

The suburban carbon sink

If you and your neighbours come to share these waterwise views and actions, your area can become a sizeable carbon sink. Your council may even become involved with an

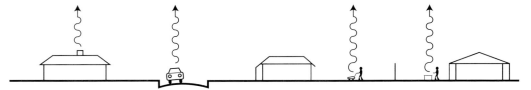

Figure 4.2a: A suburb without trees and shrubs. Most of the greenhouse gases rise, contributing to climate change.

Figure 4.2b: The carbon sink - a suburb where trees and shrubs recycle carbon dioxide.

overall strategy that includes ordinary citizens, together with professionals, planning and planting in domestic and public space. But even if this enthusiasm does not occur immediately, you will have a model garden for others to see and emulate.

A further spin-off from waterwise activities is a garden, even a street or a suburb, that invites nature to return.

The suburban aviary

Birds and other small native creatures will love your garden creation. You will have a flock of ever-changing, colourful and lively visitors who gather nectar and perhaps build their homes. Honeyeaters,

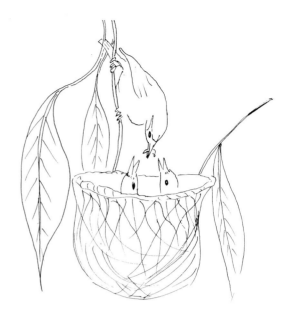

magpies and other birds of all sizes, shapes and calls will be part of your garden. Wildlife corridors could be built into a neighbourhood action plan.

Salinity

Another, perhaps unexpected, spin-off is that with more drought-proof, deep-rooted plants in suburban gardens, the chances of suburban soil salinity are reduced considerably.

Figure 4.3: Suburban seepage, which generally is unseen, is very apparent in this coastal scene.

Figure 4.4a: The natural landscape before development takes place. Forest foliage, bark, litter and roots soak up moisture. The stream is healthy.

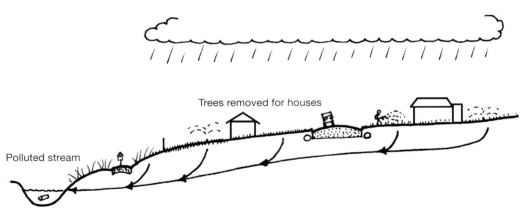

Figure 4.4b: The same landscape after development. Water penetrates the subsoil virtually unimpeded, carrying salts and nutrients downslope. The stream is polluted.

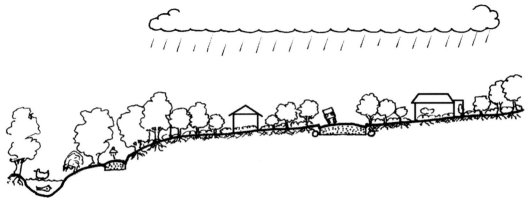

Figure 4.4c: The restored waterwise landscape emulates the original vegetation in soaking up most of the water.

Dollars and cents

For a little mental effort, some physical work and an outlay of a few hundred dollars you cannot only make dry times bearable but can add thousands of dollars to the value of your property. Add this to the fact that a well-planned garden will save on energy and water bills by creating a favourable micro-climate for your dwelling, and you realise that being waterwise makes economic sense.

A further spin-off from being waterwise is not just an enjoyable and sustainable garden but a generally improved quality of life for all involved.

Australian climate

If you examine the map and graphs in Figure 4.5 you will see that the Australian climate varies from place to place. Temperatures everywhere dip in winter but the extent of the dip varies, particularly between the south and north of the continent. Rainfall patterns are extremely variable.

There are several implications for the waterwise householder and gardener.

- You will need to tailor your design and practices to your particular climate.
- The design and capacity of water tanks you will need depends on how long your local wet and dry seasons last. You will need to consult your local water authority for the best advice.
- Vegetation types across the continent will be different because of climatic variations. The choice of plants and garden layout will vary from place to place (see also p.15).

Australian droughts

Drought and dry seasons are normal in the Australian climate. We may have become lulled into a false sense of water security because droughts occur on about a seven-year cycle. Sometimes the cycle is longer and we become complacent when droughts do not appear. But appear they certainly will and, if anything, they may become worse as the greenhouse effect takes hold.

The 2002/2003 drought is one of the most extended on record and for many Australian gardeners it has already proven a disaster.

Most of Australia has a dry season for part of the year so we all have some experience in contending with dry times. In extended periods of drought all we need do is extend that experience. The examples in Figure 4.6 represent some of our recent droughts.

The difficulty is that we generally only have water stored in our supply systems to cover seasonal dry times. When extended drought occurs supply systems become depleted and rationing occurs. If the drought is intense, water for gardens is the first restriction introduced. We have to adapt to that extreme situation.

Waterwise House & Garden

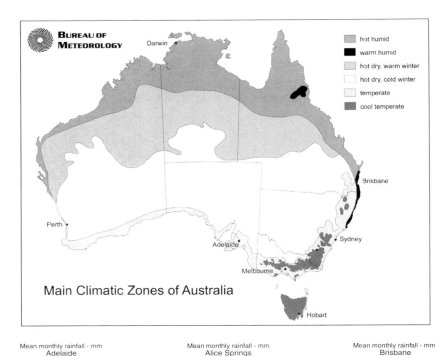

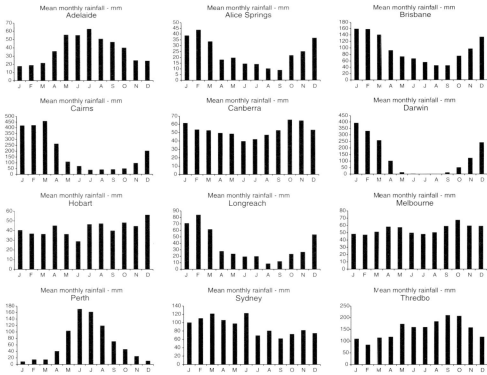

Figure 4.5: The Australian climate. **Source:** Bureau of Meteorology

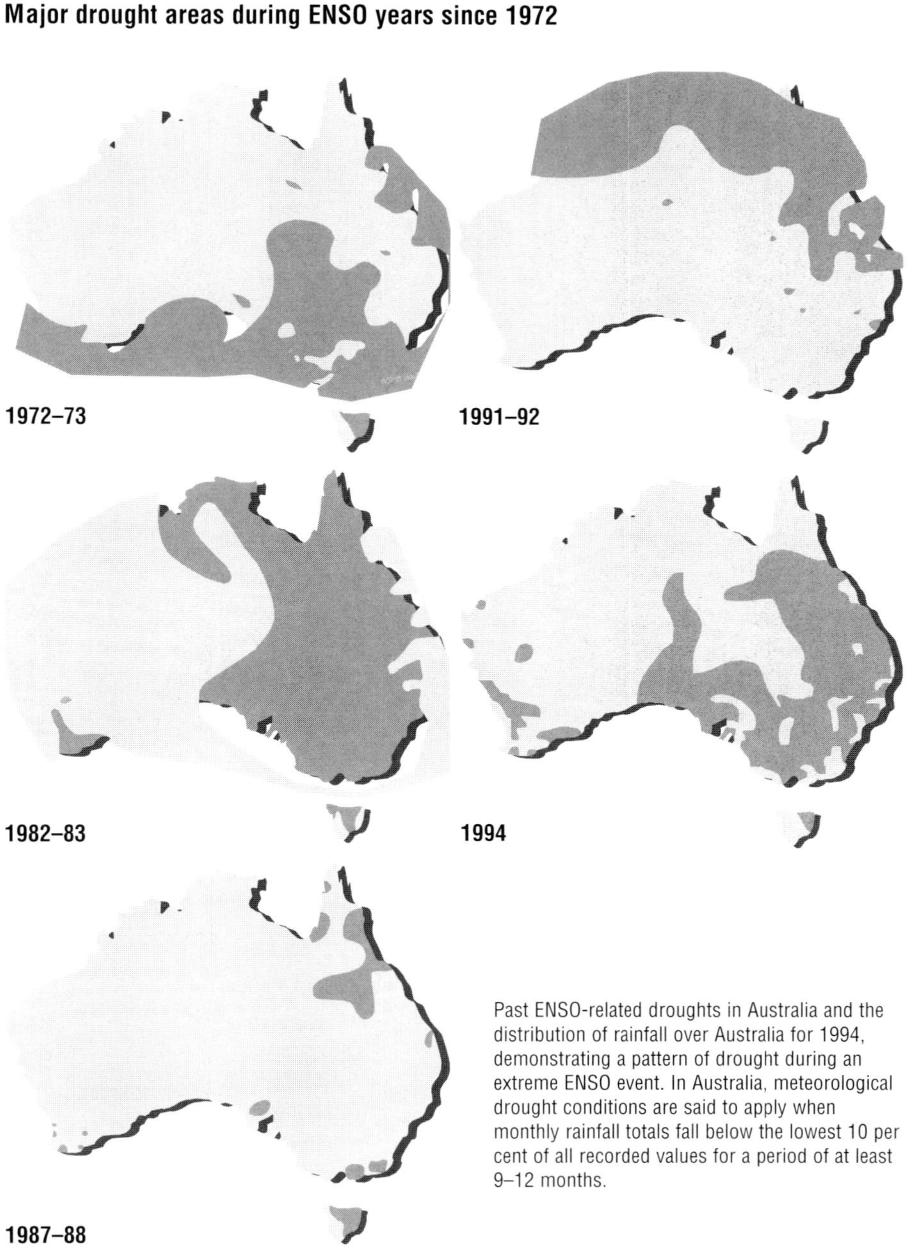

The house and garden system | 13

Major drought areas during ENSO years since 1972

1972–73

1991–92

1982–83

1994

1987–88

Past ENSO-related droughts in Australia and the distribution of rainfall over Australia for 1994, demonstrating a pattern of drought during an extreme ENSO event. In Australia, meteorological drought conditions are said to apply when monthly rainfall totals fall below the lowest 10 per cent of all recorded values for a period of at least 9–12 months.

Figure 4.6: Past major ENSO droughts. ENSO is the El Niño Southern Oscillation, a pattern of water currents in the Pacific Ocean that appears to be associated with drought in Australia and other weather extremes in the Americas. **Source:** Bureau of Meteorology

Chapter 5

Waterwise strategies

The key to being waterwise is knowing your resources and how best to use them.

Choose your approach

In developing your waterwise strategies you could adopt one of three approaches.

- Absolute precision. You can proceed as if you are running a military campaign or a corporation, with attention to the finest detail. I suspect that most householders would soon tire of this.
- Seat-of-the-pants. You can base your actions on educated guesses. This has the attraction of requiring less mental effort and achieving quicker results but, like most intuition, it can lead to error.
- Practical. Perhaps the most practical approach lies between the two extremes – not getting bogged down in detail, but being precise where it counts.

You can choose the approach that suits your personality.

Know your climate

Be familiar with your rainfall pattern and quantity. Know your wet seasons, so you can save on mains water. Know your hot dry seasons, when you will use more water on plants because of evapotranspiration.

Evapotranspiration involves losing water in the form of water vapour to the atmosphere from garden plants and other surfaces such as roofs and paved areas. Of rainfall across Australia, 90% is returned to the atmosphere as evapotranspiration. It is important to appreciate that this figure is a national average – several factors influence the rate of loss and the amount lost. For instance, if you overwater your garden the rate of evapotranspiration would remain the same as an unwatered nearby garden, but because you have added extra water the amount lost will be higher for your garden. Of course there is an upward limit to the amount of excess water that will be lost as water vapour. Any extra water after that limit is reached will either seep away or lie around in puddles, and you will be paying additional rates for the wastage.

The best source for rainfall information is the Bureau of Meteorology, which keeps records for every settled area of Australia. Try its office in your nearest capital city. Be as precise as possible in describing your location as rainfall can vary widely in relatively short distances. For instance, in an average year, where we live in a hilly area the rainfall is around 650 mm, whereas the rainfall just 15 km north on the plains is as low as 450 mm in the same year. As well, we can have wet days when people in the north miss out.

The Bureau also has records of evapotranspiration.

View the house and garden as a system

The next step is to view your whole property, from back fence to front kerb, as a total system with water inputs and losses. The inputs include rainfall and mains supply. Losses include evapotranspiration and seepage into the water table. Water usage is considered a loss unless it is recycled.

Rainfall

During the wet season in most settled areas of Australia little water other than rain is required, and often rain causes excess soil moisture. It is the dry season that requires other inputs.

If you know the rainfall seasonal pattern you have a good idea what will grow in harmony with wet/dry season cycles. The plant lists and sources of help given in the appendices will help point the way.

Rain falls on the roof of our houses and other structures, providing the potential to collect it in water tanks. If you know the area of the roof and multiply that by the annual rainfall you get a rough idea of the storage potential. I say a 'rough' idea because rainfall cycles vary across Australia. It is impractical to install tanks with sufficient capacity to hold several months rain in the tropics and subtropics, as the capacity needed would be huge. On the other hand, where rainfall is less than 300 mm a year and is spread over several months, large tanks are not required. In places with extended hot, dry seasons large tanks are a big help if rain during the wet season is sufficient.

Some sites can benefit considerably from water tanks, but some sites are not suited to their installation. Some considerations include:

- the quality of mains water
- the area of the garden. Tanks are not suited to large gardens and may be too cumbersome for small sites

- industrial pollution. Local air pollution levels may be too high for rainwater drinking purposes
- installation costs may not be reflected in the benefits
- most authorities do not allow tanks to be connected into mains supply because of potential pollution. There is nothing to stop you from making your own separate tank reticulation arrangements.

Because of these and other considerations it is best to seek advice from your local water supply company. Some water suppliers are more enthusiastic about home water tanks than others, provided there are adequate plumbing safeguards to avoid backflow into the mains supply.

You should not allow trees to overhang the roof as leaves will pollute the supply and will eventually rot your spouting. Anyway, eucalypts, wattles, pines and other trees with volatile leaves should never be close to the house in suburbs or country because of fire risk.

Unless you live in a heavily polluted area the water you collect from your roof is better drinking quality than you get from any other source. It is certainly cheaper than bottled water. Your major cost is your initial outlay for tanks and reticulation; after that you will only have minor maintenance costs.

My household has operated on tank water for 12 years and we can vouch for its superior quality and taste. We also use it for showers, baths and hand basins, and in the kitchen and laundry.

See p. 30 for details on water tanks.

Mains water

To determine your annual mains water usage, simply check your water bill. This quantity is the base against which you can measure your economic use of water in future years. Most water bills include a graph showing how current consumption compares with the previous year.

Garden evapotranspiration and seepage losses

The garden is the biggest consumer of water in an average to larger-size household. So your main waterwise concern will be to reduce the area devoted to water-loving plants and replace them with dry-tolerant, low-evapotranspiration species or completely dry areas such as paving. Much of this book is concerned with just that. See also Appendix 1.

The main water-loving plants are those in hanging baskets and pot plants, lawns, vegetables and fruit trees, and shallow-rooting shrubs and trees from high-rainfall areas of the world. Of course there are other garden considerations besides saving water and this book does not suggest, for example, cutting back on food-growing areas. What I do suggest is that, since most vegetable gardens are overwatered, water economies can be achieved by careful monitoring and control of watering practices, mulching, good design and simple technology.

A water timer is perhaps the single most effective aid to avoid the worst garden sin of all – leaving sprinklers running. A timer for each outlet will give comfort to the conscientious but absent-minded gardener. A centrally controlled timer system is more complex and costly to install, but will repay you over and over again. It forms an integral part of good waterwise garden design. The garden design comes first, followed by the sprinkler layout connected to the central control.

It is also technically possible to recycle much of the seepage water by using subsurface drainage leading to a collector sump, from which water is returned to the garden.

These technologies will be described later.

Household losses

Reducing the loss of water down the sewer is largely a matter of modifying human behaviour. Household outputs can be reduced dramatically by identifying the prime water excesses and modifying them. Sometimes this can almost be a pleasure. For instance, washing dishes three times a day can be replaced by washing them once a day. Make sure you press the toilet's half-flush button for fluids. Drainage and sewage water are discussed at length on p. 37.

Become water-conscious then a water-saver

The advice in this book makes it possible for water-conscious people to reduce water consumption by half, and even more for water-savers. In cities like Melbourne, daily residential water use is 60% of the total 480 000 ML used by the whole metropolis. If all homes became at least water-conscious, we could save 144 000 ML per day. That is a pretty sizeable dam that doesn't have to be built.

My estimate of the average Australian suburban residential usage, with the potential savings for the water-conscious and water-savers, is shown in Table 5.1. It is very informative to check the information in this table with the estimates of water tank capacities on p. 31.

Table 5.1: Estimated average Australian residential annual water consumption.

Water use	% of total	Present use (L)	Water-conscious user (L)	Water-saver user (L)
Lawns	25	75 000	37 500	25 000
Flowers, shrubs, vegetables	10	30 000	15 000	10 000
Toilet	20	60 000	30 000	20 000
Shower, bath, hand basin	25	75 000	37 500	25 000
Laundry	15	45 000	22 500	15 000
Kitchen	5	15 000	7 500	5 000
Total	100	300 000	150 000	100 000

A more detailed set of estimates is included in Appendix 7, with guidelines on conducting a water audit.

Chapter 6

Waterwise options

Choose your options according to your circumstances and inclinations. You should also note that a 'Five A' system of rating the water-saving efficiency of appliances operates across Australia. The more efficient the appliance, the more 'A's it carries on the label.

Household reticulation options

Toilet

If you look again at Table 5.1 you will see that the toilet uses 20% of the overall water supply in an average household. In a household with four people, a single-flush toilet will use over 70 000 L a year, whereas a dual-flush uses 30 000 L. Any saving will be significant.

- Install a full/half-flush cistern.
- Reduce flushing to solids only and your toilet will use only 10 000 L/pa.
- It is not a good idea to place a brick in the cistern to lower its capacity, as this can reduce the efficiency of flushing away solids and possibly damage the cistern. It may even use more water because a second flush may be required.

Shower

An ordinary showerhead in a four-person household will use around 100 000 L for five-minute showers.

- Install a water efficient showerhead and you have reduced that usage to 50 000 L/pa.

- Reduce shower times to half or less and reduce the usage to 25 000 L/pa. Most people spend far too long in the shower. Time yourself to see how long you take. People in my house seldom spend longer than three minutes in the shower, because we use only tank water and are very careful not to waste it.
- When you first turn the shower on, place a bucket to catch the initial cold water – there is always a delay until the water becomes hot, a delay that can waste many litres. You can use this water on pot plants.

Washing machine

Four washes per week in a large washing machine will use 40 000 L/pa.

- Reduce the number of washes by half and the usage is down to 20 000 L.
- When purchasing a new machine consider a front-loading model which can use a little as 50 L per cycle; top-loading models may use 180 L per cycle.
- Whenever possible wash only with full loads.
- Redirect spent water onto the garden. Caution is needed with the type of detergent used.

Laundry sink

Place a bucket under the tap to capture the initial cold water while waiting for it to become hot. The water may be used for any hand-washing of small items or on the garden.

Hand basin

Four people using a hand basin twice a day for three minutes will use 8000 L/pa.

- Use sensible economies for cleaning teeth, shaving and washing hands and face and that figure can drop to 2000 L/pa.
- Redirect spent water onto the garden.

Kitchen sink and dishwasher

Hand-washing twice daily can use 10 000 L/pa.

- Use a dishwasher which is water-efficient and that figure can be reduced to 7000 L/pa.
- Redirect spent water onto the garden.
- Wash up once daily and if using a dishwasher, do so only with a full load.

Bath

- Reduce the depth of water used.
- Bail with a bucket if water quality allows use on the garden.

Garden watering methods and appliances

The waterwise guiding principle for watering the garden is to ensure that each plant gets enough water for healthy growth, but no more. Water should be evenly spread throughout the root system.

Aim to moisten the root zone of plants. Moisture on leaves or stems can encourage fungal diseases. Watering leaves in hot sunlight may cause the leaves to cook when the sun heats the water left behind and rapid changes in leaf temperature damage plant cell stability. Conversely, some plants can be helped by foliar watering on frosty mornings. This thaws them gently before the sunlight strikes and thaws them too quickly, bursting the cells and resulting in the characteristic burned appearance of 'frost-burn'.

Overhead watering can encourage weeds and is subject to misdirection by wind. It is also less efficient in penetrating thick, less porous mulches.

Since most people are familiar with many of the garden water distribution appliances and fittings, not all have been described here in detail. Most are described fully in manufacturers' brochures, which include flow and coverage details. What the manufacturers have not included are the shortcomings of their products.

If you are having your irrigation system installed by a contractor, be present on the day of installation to ensure no corners are cut.

Bucket using mains water

A bucket placed under a tap or shower will collect clean water if it is done before soap is applied.

Buckets are only efficient for watering individual plants in the garden or in pots, unless you have a tiny lawn. A bucket of recycled water used every day can save 1800 L/pa. To save your back, try to water plants closest to the house.

Caution is needed with any detergents (see p. 39).

Hose using mains or tank water

Watering by hose is not the most efficient means of irrigation and it can waste your time, even if it makes you feel good. Once-a-day watering for 30 minutes can use 60 000 L for the summer period. Watering once a week for the same time uses only 8400 L for the summer period.

If you are forgetful, use a timer on the tap serving the hose. Judge your watering times based on plant type, soil type and micro-climate.

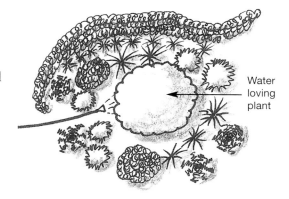

Figure 6.1: Hoses are handy for irrigating specific water-loving plants isolated within an otherwise dry zone.

Hoses are handy for:
- spraying soapy water off cars after they have been washed by bucket. Wash the car on the lawn if possible
- vegetable stress relief when, for instance, a short sprinkle at the end of a hot day gives them a lift
- directing a fine spray onto an area which does not justify installation of pop-up sprinklers
- once a month or less watering of established moderately water-loving trees or shrubs isolated in otherwise dry areas.

Sprinklers using filtered mains water

A well-designed sprinkler system is the best waterwise means of regulating and distributing water to garden wet zones. The key is an effective timing mechanism that allows

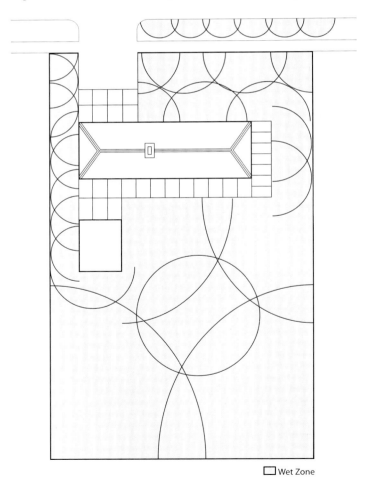

☐ Wet Zone

Figure 6.2a: A sprinkler-seller's fantasy, in which every square metre of garden is more than adequately covered by sprinklers.

you to control rates of water application. The design and type of plants selected for traditional gardens with lawns, flowerbeds, water-loving shrubs and trees is the main cause of overusing water. The fact that sprinklers are used in such gardens does not mean the sprinklers themselves are inefficient.

Sprinklers are efficient because:

- they can be located to water-specific 'wet areas' such as lawns
- they can, and should be, controlled by timers
- they provide the opportunity to measure water application quantities.

In summer, watering the whole area in an average garden twice daily for 30 minutes would use 180 000 L for the season. Using the same sprinklers once every three days for 30 minutes would use 27 000 L.

Figure 6.2b: Waterwise design for the same property with limited wet areas which require one-third the water usage.

Sprinkler selection
Ensure that the sprinkler suits the garden situation.

- Sprinklers should not spray roads or other paved areas.
- Choose a 360°, 180° or 90° spray pattern to match the location.
- Choose larger-droplet types for wind-affected situations to avoid misdirected sprays.
- Schedule your control timer to accord with the application rate of individual sprinklers.
- Make sure that similar-type sprinklers are on the same pipeline and timer control.

Waterwise sprinkler system design
Good design, that limits wet areas but still gives quality of life, can cut water use. The same area can look as good, if not better, as shown in Figure 6.2b. Design your garden first, then use the manufacturer's guidelines for sprinkler coverage of wet areas. Your supplier should provide a manufacturer's guidelines. A good supplier may provide you with a design free of charge.

Professional advice
When seeking professional advice it pays to:

- check that the designer has qualifications and is a member of the Irrigation Association of Australia, a voluntary organisation that can help with conflict resolution
- obtain a written quote
- check the plan against the advice given in this book.

Misters using filtered mains water

A mister is the smallest form of sprinkler. Misters are suited to hanging baskets, especially ferns, large pots or tubs, narrow flower and shrub beds and some vegetable types, where the mulch is porous.

Advantages

- They can deliver water to confined areas.
- They are easy to install and remove.
- The controlled-flow types can deliver a measurable amount of water.
- They can be set on different height risers to suit plant height.
- They can be laid out in a flexible manner for many different designs.

Disadvantages

- They are associated with plant fungal problems.
- It is difficult to penetrate thick non-porous mulches.
- Delivery pipes can be easily cut accidentally.

- They tend to clog unless water is filtered.
- Insects or recycled water can clog nozzles.

Porous (sweat) hoses using filtered low-pressure mains or tank supply

This is an excellent product, made from recycled rubber, that works well if it is installed correctly. Porous hoses can be used under mulch and even just below the soil surface (about 5 cm deep) so that direct evapotranspiration and ultraviolet light damage are minimised. They are best used for wet-zone plants such as flowers, vegetables, fruit trees and other water-loving plants.

Figure 6.3: The sponge-like hose allows drops to be distributed evenly along its length.

Advantages

- They distribute water near the root zone.
- The water is distributed evenly along the hose length.
- Roots seldom penetrate them. If roots might be a problem, place hoses on the soil surface and cover with mulch.
- Weed growth is reduced.
- Under mulch, their use is not affected by sunlight or wind so they can operate during the daytime.
- They are ideal for long wet zones.

Porous hoses operate best at low pressure so you are advised to use a pressure regulator set at around 60 Kpa (8 psi) at the inlet end of the sweat hose distribution system. Ideally you should have a 200 mesh filter if a master filter is not already installed at the entry to your garden system. Also, it pays to use a flow regulator to keep the flow down to 1.3 L/min per 10 m of hose. Excess pressure considerably shortens the life of porous hose.

Disadvantages

- Like any buried system, they can fall victim to the garden fork or spade of even the most alert gardener. The solution is to mark the lines with a stake or a rock or brick.
- They are recommended only for fully treated and filtered recycled water.
- Buried systems cannot be observed to see if they are working and a moisture probe will be needed to test soil conditions.

Dripper hoses using filtered mains or tank water

Dripper hoses come as a unit, with the drippers built into the line by the manufacturer. They are versatile and can be used in similar situations to porous hoses. The difference is that the hose is not porous; it has holes at intervals where plants can be placed alongside to benefit from water drips. The intervals are from 200 mm upward. Pressure and filtration needs are similar to those of sweat hoses.

Advantages

- The ultraviolet-treated line is best placed on the surface of a suitable mulch so that you can inspect to see if it is working. The concentrated drip will penetrate most mulch types.
- Although it is probably best to water at night, you can use dripper hoses early or late in the day when evapotranspiration is low.
- Well-designed types normally do not become clogged with insects.
- There are different spacings of drip holes for different crops.

Disadvantages

- They can be cut or pierced if covered by mulch, but this is less likely if lines are on the surface.
- Flow quantities are hard to calculate, so a moisture meter is needed to determine soil moisture levels for different periods of time.

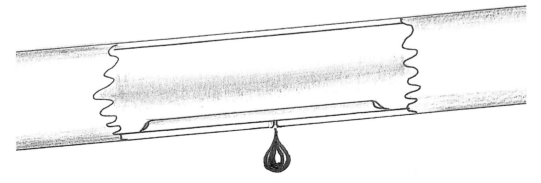

Figure 6.4: An internal sleeve allows water to soak and drip out but does not allow insects in. The hose has a longer life than the sweat hose.

Individual drippers using filtered mains or tank water

Drippers are ideal for hanging baskets and some pots. They come in several styles.

Some adjustable drippers operate like a cross between a dripper and a small sprinkler. They can be useful for isolated water-loving plants such as fruit trees in otherwise dry zones or espalier situations. Some manufacturers recommend that you use adjustable drippers in-line in vegetable rows with a dripper for each vegetable. This could be expensive and require a lot of attention adjusting flows on each dripper.

Figure 6.5: Individual dripper for a hanging basket.

Filters for mains supply or tank water

Filters range in size from single tap fittings to whole-of-household and garden systems. The filters discussed in this book are designed to keep fine solids from entering the pipeline supply and clogging sprinklers and drippers. They do not filter out dissolved chemicals.

The finer the filter, the more impediment there is to flow. If you live in an area with low water pressure you are advised to be extra careful in your choice of filter.

A 200-mesh filter should protect all fine sprinkler outlets.

Figure 6.6a: A single tap filter contained within the tubular housing.

Figure 6.6b: A total system filter. Care should be used when cleaning these filters to avoid damage. Lightly scrub the filter screen with a toothbrush.

Figure 6.7a: A single-tap timer.

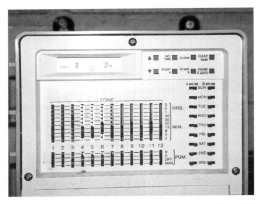

Figure 6.7b: A total-system timer controlling 12 stations. Each station's watering time and quantity can be individually set.

Timers

Timers are a definite benefit in any waterwise system. They come in a range of sizes and complexity; from individual tap timers to whole-of-system timers at a single station.

Each outlet in an electronic system has a solenoid control, which is connected by fine wires to a central switchboard timer. It is important that these wires are well insulated from soil chemicals and physical disturbances, because you will have endless trouble chasing frayed underground wiring and, worse still, installing new underground wiring. For peace of mind, pay the extra for quality wiring.

Simple single-tap timers are best for the average garden, because they are less costly to purchase and maintain than the total-system models. Zones differ in watering requirements. Some garden zones such as fruit trees may require watering only once a week, other areas may require daily or no watering. These variations make it difficult to set total-system timer controls to serve all needs.

A good idea is to take your moisture meter and test the soil around the outlet over time intervals to determine which timer setting you need for your soil type to promote healthy plant growth without waste (see p. 49). This will save you the trouble of calculating flows first from each outlet, then totalling them to determine the flow for the whole system.

Moisture meters

Moisture meters come in varying complexity, from the simple hand-held gardener's probe to the automated models used in commercial nurseries and horticulture. The hand-held model is adequate for most garden situations.

When using the probe make sure that you drive it to two-thirds of its full depth in the soil beside the plant being monitored. Take care not to pierce any buried pipelines or roots. If you are using a timer distribution system you will only need the probe for initial testing to determine how long the outlet should be left on for the right amount of water to the plants in that part of the garden.

Good meters have a numbered dial and instructions on plant-watering values. For example, the model in Figure 6.8 shows that a reading of 4 indicates a water-loving plant, 3 indicates an average plant, 2 indicates a semi-hardy plant and 1 indicates a hardier plant.

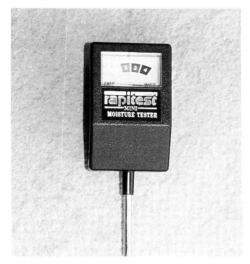

Figure 6.8: A simple probe suited to most home gardens.

Automated systems using mains pressure systems

An automatic system is the most efficient means of delivering water to all wet areas of your garden. An automated system package includes:

- a control panel with a transformer connected to your power supply (best located undercover in a central accessible location)
- wiring to all solenoid control valves
- pipes for water to run in. Most manuals recommend 13 mm poly pipe for fine sprinklers and drippers, and 19 mm for pop-ups. I favour 19 mm pipe to all outlets, because the extra flow to each point gives an even spread to all outlets along the total length
- many different fittings for joints and valves.

It is best to go to a supplier who can help with a layout plan to match your garden design. They should provide the plan free of charge. If they do not, go to someone who does. Some supplier designers tend toward overdesign, to avoid constant maintenance.

Pot plants

Pot plants are increasingly being used in units or properties with small gardens, because they save space and are attractive. Pot-plant gardening can be the most or the least water-efficient gardening practice, depending on the way it is done. There is nothing worse than a poorly managed pot plant.

The following rules will assure success.

- Paint terracotta pots inside with a sealant to avoid the plant drying out, or keep the plant in a plastic pot inside the terracotta pot disguised by a covering of mulch.

Figure 6.9: A catching-dish under a pot plant for recycling water.

- Use a good-quality potting mix that will hold the moisture. Although worm castings are superior soil improvers and water retainers it is not advisable to use them in pots. They may contain worm capsules which will hatch into young worms that will eat their way through the soil's organic matter, leaving the pot to dry out quickly.
- Choose pots to suit the size of the plant. Keep in mind the root growth pattern of the plant when selecting the pot. Of course, if your space is limited there may be some plants you grow in pots to inhibit their otherwise excessive growth. For instance, a bay tree in a large pot will not grow into the huge specimen it would if planted in the garden.
- Use a moisture meter or the one-finger test to see if the plant needs watering.
- Develop a watering routine.
- Place the pots where you will see them daily so that you will notice any moisture stress and be reminded to give them a drink.
- Place a time-controlled misting system around your pots if you spend time away.
- Place the pots within easy walking distance of the house if you water by bucket.
- Recycle any excess drained water by placing a catching-dish under the pot.

On-site catchment and storage in water tanks

Water that falls on the roofs of houses and other buildings is a huge but untapped suburban resource.

Regulations

Water authorities differ in attitude towards tank water, ranging from strongly negative to highly enthusiastic. My attitude is that the use of water tanks depends entirely on your situation and your resources. I can enthusiastically endorse it. For my household, tank water supply in the Central Highlands of Victoria is better than any reticulated city supply.

Councils, EPAs and water supply authorities are very wary of legal issues associated with approvals because of health considerations and other hazards. They may object to a water tank on the following grounds.

- The installation is unsafe.
- Overflows are likely to cause flooding and undermine structures.
- The tank may be a source of health problems, such as a breeding area for mosquitoes or contamination from the roof.
- The tank contravenes building regulations, for example it is too close to boundaries or blocks the neighbours' view:

However, if you install and manage your tank system properly, you should have no problems.

Uses

You should at least be able to use your roof supply for washing clothes or mixing with grey water. ('Grey' water is all water that has been used domestically, such as in the kitchen, laundry and bathroom – except for the toilet. Water from a toilet is known as 'black' water and cannot generally be recycled domestically.) In very small gardens you could use tank water on your vegetables and fruit trees. A tank can be an excellent standby in drought. Another use for tank water, even in polluted areas, is to supply the toilet cistern. This immediately saves 20% of your water supply.

Capacity

The beauty of the roof as a rainfall collector is that runoff is 100%. If you know the rainfall you can calculate the likely tank capacity. Simply multiply the roof area by the maximum concentration of rainfall over two months of your wettest season. In the tropics or other wet regions allow for a longer wet season. You have to allow for your consumption for the whole year. If there are, say, six members in the household you will consume much more than a household with two members. Conversely, if two people have to go through six months of dry weather they may consume more than a six-member household with a dry season only two months long. If you don't want to work out your requirements, you can consult your local water authority instead.

An average roof in Melbourne can supply around 80 000 L a year. In Sydney and Hobart that figure could be as high as 100 000 L, and in Brisbane around 150 000 L. Table 5.1 (p. 18) shows that many Australian households could be self-sufficient in water supply if they installed water tanks and became water-savers.

If your budget does not permit you to purchase a large tank system, try a small one to start with and add others as your financial situation allows. On-ground tanks are the cheapest to install, but require a pressure pump. One trick is to hybridise the system by adding a small pressure tank above the height of the ground-level tank and connect them using a small solar pump if you have adequate sunshine year-round. In this way you can have considerable storage capacity at ground level and a gravity-fed household supply.

To determine the optimum water tank capacity to suit your roof area and the local rainfall distribution throughout the year you can use a complex formula, which should be available from your local water authority. If you are less mathematically inclined ask your local tank supplier and check their recommendations with the local water authority. Those in the industry know the local requirements.

For instance, a 200 m^2 house in Melbourne could use a 60 000 L capacity tank to serve a household of four with drinking water, shower, laundry, and kitchen supply. If the water is used for flushing toilets only, the tank size could be reduced to 20 000 L or even less, if residents use water-saving economies. If tank water is to be used on the garden as well, the size should be increased, budget permitting. See also p. 11.

Drinking water

If you wish to use tank water for drinking, showering and bathing, particular conditions must be met.

- The roof must be metal, as tiles or slates can harbour fungi and bacteria, which can harm anyone who drinks the runoff. Colourbond or other special coating is suitable.
- Dwellings should be away from major roads and windborne industrial pollution that lodges on roofs. You can boil tank water to kill organisms if you are concerned about any contamination. Boiling will not get rid of metallic or other airborne pollutants, but it will kill most pathogenic organisms. Boiled water is best shaken and left overnight to oxygenate it before drinking. If you are confident that there are no airborne pathogens, boiling is not necessary.
- The area should be free of birds such as seagulls or pigeons that are likely to gather – and poo – on your roof.
- Roofs should not have overhanging trees that will drop leaves in the spouting. Some eucalypt leaves can contain large amounts of tannin.
- The tanks should exclude light to discourage algal growth.

Installation

Make sure there is adequate access to the tank site for installation. Suppliers may just drop the tank off for you to deal with if the access is poor.

When installing your system, make sure that you have a means of diverting the first flush of water off the roof after a long dry spell. This is to get rid of excess dust or leaves that tend to collect on any roof. Your plumber can advise which type of diverter suits your type of roof construction. Gutter contamination by leaves is one reason that many councils oppose tanks. Gutter guards, available from hardware stores, save cleaning out spouting. The best way to remove gross contaminants such as leaves, finer grit or other solids which may lodge on the roof or spout is to install the first-flush water-diverter on the downpipe or at the top of the tank. Manufacturers of good rainwater tanks should supply such diverters for tanks as standard options.

If you wish to filter the water even further for fine dust, rust and other sediments that can affect taste, install filters in-line or at taps.

All such accessories should be displayed in manufacturers' brochures.

Make sure you read the manufacturer's installation instructions thoroughly, so that you know that whoever installs your tank does it properly.

Tank construction

The first step is to choose the type of tank: metal, concrete or polyethylene. Metal tanks have been in use for well over 100 years. Metal includes galvanised iron or steel, zinc-finish aluminium, or stainless steel. Galvanised tanks last longer if coated on the interior with a polymer lining, and need a wooden stand rather than a metal one that may react chemically with the tank. Concrete tanks are regarded as permanent. Polyethylene tanks are lighter and cheaper to purchase and install. Each type of tank has its application.

Figure 6.10: The wide range of tanks available. **Source:** Photo courtesy of Polymaster

Inspection and maintenance

Make sure the tank is provided with an inspection hole that is insect-proof so you can keep out mosquitoes between inspections. Make sure that children cannot enter the opening and, if you have to enter to inspect the tank, make sure you have breathing apparatus and someone is nearby with extra breathing apparatus. Gases in the apparently empty tank can overcome you. Never try to rescue anyone affected by gas unless you are wearing a breathing apparatus. Working in confined spaces can be dangerous

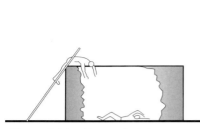

Figure 6.11a: Never climb into a water tank unless you are wearing a breathing apparatus.

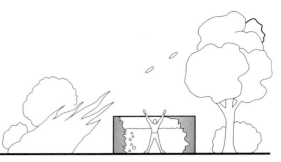

Figure 6.11b: If you live in a bushfire area do not use your tank as a refuge during a fire – the water can boil and so will you.

and is subject to government regulations, e.g. *Australia/New Zealand Standard: Safe Working in a Confined Space AS/NZ2865:2001* and *Occupational Health and Safety (Confined Space) Regulations 1996 (Vic)*.

Hire contractors to clean out leaves, as they will have the right vacuum equipment. Do not clean off the film that builds up on the inside of galvanised tanks. This natural coating is actually protective and reduces the risk of corrosion.

Elevated water tank

Water tanks on stands are suitable for gravity-fed outlets such as a kitchen drinking-water tap. Conventional stands are not the most attractive piece of garden furniture, but a little imaginative design can work wonders. For instance, you can convert the stand into a small shade house, tool cupboard or, as I am constructing at the moment behind my potting shed, a poultry shed. You are limited only by your budget and your imagination. Storage capacity is considerable in an elevated tank, but is generally not enough to supply more than kitchen water. Polyethylene stainless steel or galvanised tanks can be used.

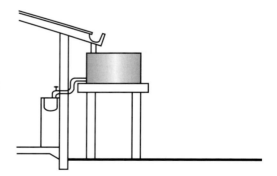

Figure 6.12: A low-pressure above-ground tank connected directly to the kitchen with no connection to the mains supply.

You will probably need separate plumbing to the kitchen as regulations do not allow direct connection from tanks into your mains supply due to potential contamination by dirty tank water.

Your water authority can advise your plumber how to connect into mains supply without the risk of contamination.

On-ground water tank

An on-ground tank is the least expensive to install but it does have some shortcomings. As implied above, it presents a pressure supply problem. The tank is not placed actually on the ground but on a very low stand, concrete pad or bed of sand, so it needs a small elevated pressure tank as described on p. 34, or a supply pressure pump. A 100 mm thick concrete slab or a similar thickness of sand forms a satisfactory base for an on-ground tank. Make sure the base extends beyond the edge of the tank to form a lip edge.

Another problem is its appearance, but this can be overcome with a little architectural imagination.

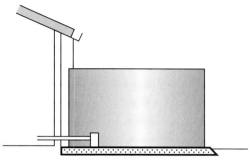

Figure 6.13: On-ground water tanks require a firm base so that they will not tilt or be subject to pressure from cracking foundations.

Space also presents a problem, as tanks do occupy a considerable area. In a small or established garden this could be awkward, if not impossible.

They can be made of any of the materials mentioned earlier.

Plastic tanks are more manoeuvrable and easier to install. You may like to run a flush of water through before you drink from it the first time. You can use this first fill on the garden. Plastic tanks come in a range of environmental colours.

Concrete tanks have the advantage of lending themselves to decoration or masking with rough finish to make them fit into the landscape. They are durable but need mechanical lifters to manoeuvre them during installation. Run an initial flush of water to wash any lime leachates out. Do not use that water on acid-loving plants. Concrete will not burn but it can be damaged by close intense fire, so keep any potentially flammable vegetation well away.

Your swimming pool

Even if you do not install a water tank you can use your roof supply to supplement your pool supply by redirecting your downpipes into the pool.

Underground water tank

It is possible to store all your runoff in an underground water tank, but this requires a major excavation and construction operation suited only to the most prosperous and dedicated water-savers. It could also mean a considerable alteration to your landscape design. However, if the tank is large enough you will end up with an almost unlimited supply of rainwater for household purposes.

You can install a precast concrete tank or an on-site cast construction.

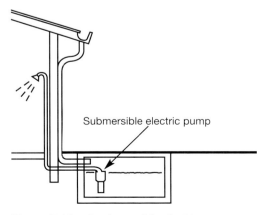

Figure 6.14a: A submersible electric pump can provide mains pressure.

Figure 6.14b: A landscape treatment of the top of a buried concrete tank.

It is advisable to seek professional advice before installing an underground tank. Take care to avoid damaging underground services during excavations. It is possible to construct an attractive garden feature to disguise the underground tank.

The tabletop in Figure 6.14b acts as an inspection hatch lid. Ensure the paving slopes slightly away from the hatch for drainage. The hatch base and sides should be waterproofed to avoid any possible contamination.

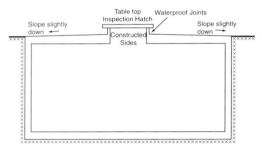

Figure 6.14c: Cross-section of a buried tank.

Tank water for toilet cisterns

It is possible to connect your tank to your toilet cistern and/or garden, so that even if the water is not drinkable it still can be used productively to make major water savings. This will mean additional plumbing and the installation of a dual-check valve so that the rainwater does not mix with the mains supply. It can be argued that tank water in a polluted area adds no more contaminants to the garden than polluted rain falling directly on the garden or, indeed, air pollution acting on its own.

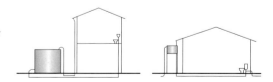

Figure 6.15: Single-storey and dual-storey house applications. The dual-storey application will require a pump.

Ponds

On larger properties ponds can be filled from the roof and pond water may be used for summer irrigation, especially in drought. On rural properties dams will serve this purpose, but roof water, being so precious, is more likely to go into a tank. Overflows can feed into the dam.

The dam at my house serves a number of functions:

- It provides a wildlife habitat.
- It is an emergency source of water supply for the garden.
- It can be used as a fire-fighting supply.
- It provides a fire buffer zone.
- It provides a restful outlook from the house.

This dam is filled by natural catchment and tank water overflow.

Figure 6.16: Dam on the north side of the author's house.

Recycling household wastewater

Recycling is a practice every responsible citizen is keen to undertake. Recycling all domestic wastes on-site will become a major issue this century because of the escalating costs of transporting and treating bulk wastes in sewerage systems. Recycling domestic wastewater is no exception to this prediction.

There are costs and rewards with reuse of household wastewater. The rewards are easy to identify – several are mentioned in Chapter 2, 'The value of water'. The costs can be measured in time, money, environmental protection and technology. For example, if you live in a very wet climate there will be additional costs for large storage tanks during the wet season. I should warn you that recycling most household water is not a simple matter. Your efforts at water savings at home can be better directed first to waterwise practices, collecting rainwater and recycling garden wastewater. For a background discussion see Chapter 12.

You should obtain the latest regulations or advice on domestic recycling from your state's Environmental Protection Authority. New South Wales, for example, has very high standards.

There are two classes of wastewater, known as black water and grey water.

Black water

Black water comes from the toilet and consists of solid wastes and urine diluted by flushing water. This water should never be recycled unless it is thoroughly treated by highly technical methods, which are generally considered too costly by suburban homeowners.

It is technically feasible to reuse both the solids and liquids. The Chinese and many other cultures have been doing it for thousands of years to sustain their agriculture. Unfortunately, disease is associated with the use of black water and many authorities discourage its use even when treated, because of this possibility.

Probably worse is the array of chemicals that now pollute such waste, in particular heavy metals. Nationally backed scientific research is needed to develop an economic recycling system for every household. Such technology would save millions of dollars by doing away with the necessity of providing new sewerage systems for our ever-expanding suburbs.

The best technology I have seen was the 'Dowmus' composting toilet detailed in my books *Worms Downunder Downunder* and *Worm Farming Made Simple*. You will find an advanced Dowmus model in *Sustainable House* by Michael Mobbs, which also contains a lot of useful information on sustainable water supply, home design and living. Sadly, Dowmus is no longer in business. This is most unfortunate because its household waste recycling technology was affordable and fairly simple. It also had the benefit of recycling most organic waste likely to be produced in the average home, resulting in a very useful and safe soil conditioner. I have no doubt that this technology will reappear, even if under another name.

It is vital that no harsh household chemicals enter a biological system such as the Dowmus, as the chemicals may kill the bacteria and other organisms that do the good work of treating waste products.

Grey water

Grey water comes from baths, showers, hand basins, washing machines and kitchen sinks. It comprises water containing soaps, detergents and some contaminants such as heavy metals, like the zinc used in cosmetics. Other contaminants include aluminium found in deodorants, boron found in some detergents, caustic soda in drain cleaners, detergent enzymes used to break down fats in dishwashing water, oils such as eucalyptus in concentrates, phosphorus included in many detergents, salts forming part of most detergents especially laundry concentrates, and washing bleaches. Some occur in concentrations up to 10 times that considered safe for use in the garden.

Grey water also contains lint from washed clothes and other fabrics, as well as faecal matter from our bodies and nappies.

People being treated for cancer may pass some radioactive or chemotherapeutic drugs in shower or bathwater. These substances are damaging to both soil and plants.

Micro-organisms abound in a frightening array. They include disease-carrying viruses, bacteria, protozoa and other micro-organisms, which can cause a range of diseases such as gastroenteritis, meningitis, respiratory disorders and tetanus.

While it is technically possible to recycle shower and bath grey water for use in the toilet cistern, unfortunately it is not allowed. Shower and bath wastewater can contain faecal matter. Flushing water gives off contaminated aerosols, which you can breathe in and risk disease.

Do not allow grey or black water to contact skin or your normal drinking water.

Recycling untreated grey water onto the garden

You should know the implications of using untreated grey water on the garden so that you can make your own judgement.

- As grey water can contain substances such as fats and lint it can clog valves, drippers, misters, perforated hoses and any appliance with fine openings.
- It will eventually clog soil pores, blocking air, water and root movement.
- Detergents and soaps contain substances such as sodium that will burn foliage if grey water is applied overhead.
- Grey water is inclined to be alkaline so it should not be applied to acid-loving plants such as citrus. It can also corrode appliances.
- Grey water may contaminate properties downslope from yours, with legal implications.
- Grey water can kill young plants and contaminate food crops, especially if the water is hot.
- The smell and appearance of grey water may offend you or your neighbours.

There are traditional common-sense solutions to these problems, but difficulties may remain.

- Use a grease trap.
- Run the grey water through a cloth bag to collect lint and fat. Rinse and dry the bag periodically.
- Rotate the application of grey water around the garden.
- Limit the household use of bleaches, disinfectants, detergents and lanoline-laden soaps.
- Apply grey water over thick mulch that can absorb and filter harmful solids so they can decompose over time.
- Apply it to flat areas and limit its use on slopes.
- Use it around established ornamentals, not food plants.

Washing machine and washing tub grey water

Washing machines are heavy users of water. The quality of the water may be all right for plants for one summer only during drought, provided that harsh detergents or bleaches are not used. There are a few brands available that are low in phosphorus and sodium, as well as aluminium, boron, enzymes and fats.

Keep in mind that grey water from washing machines can contain harmful micro-organisms from washing clothes such as underwear or nappies, which may contain faecal matter. Although faecal particles in grey water are more minute than in black water, avoid direct contact with the skin, mouth or nose.

Lint in washing wastewater is another problem if applied to the soil. The small fibres tend to combine with other substances to clog soil pores, virtually killing it. Some of these fibres are artificial and will not break down rapidly, if at all.

You should also remember that any contaminants could flow into neighbouring land via surface or subsurface waters.

The washing machine is generally attached to a laundry trough, so a hose can be attached to the drainpipe for emergency use during drought. Avoid using the water on vegetables, especially root crops.

A clean plastic garbage bin can be used for temporary storage of washing machine overflow immediately behind the laundry wall, but you should never store grey water for more than 24 hours, because it is subject to problems such as algal build-up.

The problems involved in reusing grey water should make you conscious of the environmental impact of washing, cleaning and strong chemical 'germ-killers' which, in a standard sewered system, we don't think about when they disappear down the sink after use. Reusing grey water forces us to reconsider the value of these products.

Shower: clear or grey water

Showers are prolific users of water. You can collect some for the garden by placing a bucket under the shower while waiting for the hot water to flow. This is clear, not grey water. You can even use it as a disciplinary timer to conserve water use while having a shower. When the bucket is full, you should have finished your shower.

If the shower is elevated above ground level you may be able to tap into the drainage system. This is grey water and should only be used during drought periods. It can contain detergents, disinfectants, soap and bleaches. The continued use of grey water can destroy soil and plant health. It will almost certainly contain faecal matter, so don't use it on vegetables.

Bath: grey water

Baths can be more water-economical than showers and you should be able to bail at least two buckets full. If the bath is elevated above ground level you may be able to tap into the drainage system. Again, this is grey water and should only be used during drought periods.

Hand-basin: grey water

Hand basin drainpipes can also be tapped. This is grey water and should only be used during drought periods, following the precautions listed above. This water can be piped to strategic garden locations but avoid its use on vegetables.

The good news is that you can buy a hand-basin/toilet set in which the hand basin feeds wastewater directly into the toilet cistern. The only contaminant in the hand basin is soap, so it is not considered a health risk. The bad news is that this technology, which is extremely popular in Japan, has yet to be released in Australia.

Kitchen sink or dishwasher: grey water

Figure 6.17: Hand-basin toilet. Source: Corona

Kitchen wastewater is generally not reused because of powerful detergents and fat from greasy water. Although it is unlikely to contain faecal matter, it can contain most of the types of chemical contaminants that showers and baths do. The chemicals may even be harsher. Avoid its use on vegetable beds and the rest of the garden even during drought.

There is one exception in the kitchen. If you do not use salt in cooking, drain the cooking water into a jug and apply its nutrient-rich contents to your pot plants or hanging baskets after it has cooled.

Septic tank effluent: black water

You may be tempted to use untreated septic effluent on the garden. Unfortunately, the effluent has a high pathogen content and can be dangerous. You can grow shallow-rooting plants such as citrus over effluent lines.

Treatment and reuse of grey water

The *Model Guidelines for Domestic Grey Water Reuse for Australia* defines grey water as wastewater from bathtubs, showers, bathroom wash basins, washing machines, and laundry tubs. It excludes untreated household sewage, wastewater from kitchen sinks, dishwashers, garbage disposal units, laundry water from soiled nappies or wash water from domestic animals. This further segregation of grey waters complicates the plumbing alterations and household practices required for an approved recycling system.

The *Model Guidelines* lay down construction and operation standards intended to prevent system blockages, leaks, overflows, foul odours, over-irrigation of gardens, contamination of drinking and washing water, damage to the environment, and general plumbing malfunctions, while delivering the water to where it can be reused.

Health and safety requirements cover the need to:

- direct any faecally contaminated water to sewers
- install backflow devices to avoid contamination of mains supply
- install overflows to the sewer
- colour-code pipework and fittings
- maintain grey water operating pressures below mains pressure
- avoid contact with humans
- avoid using laundry or animal wash wastewaters
- avoid applying the treated water where it can reach land surfaces, waterways, domestic supplies or stormwater drains
- avoid its use on food crops
- signpost irrigation areas with a sign reading 'RECYCLED WATER IRRIGATION AREA. DO NOT DRINK. AVOID CONTACT'.

Application restrictions require that:

- grey water cannot be used for toilet flushing
- it must be applied above ground
- absorption trenches such as those used for septic effluent not be used
- no evapotranspiration trenches are to be used
- no unapproved use be made of swimming pool and spa overflow water.

The local water authority will require:

- a permit
- anyone buying a house with an existing grey water system to reapply for the permit (which is usually considered a turnoff by potential purchasers)
- the installation of connections to plumbing be made by a licensed plumber
- the need to meet Australian standards and be approved by the local authority
- the system must be test-run before approval.

Reuse systems
Design requirements are as follows.

- The system must cater for a minimum of five persons per household even if there are not five people living there. Flows must be calculated on a per person basis for the bathroom and laundry.
- The system must be located according to prevailing winds, nearby buildings, ground slope and soil types, and must be based on a thorough site evaluation.

Typical designs
The following designs show the basics in such a reuse system.

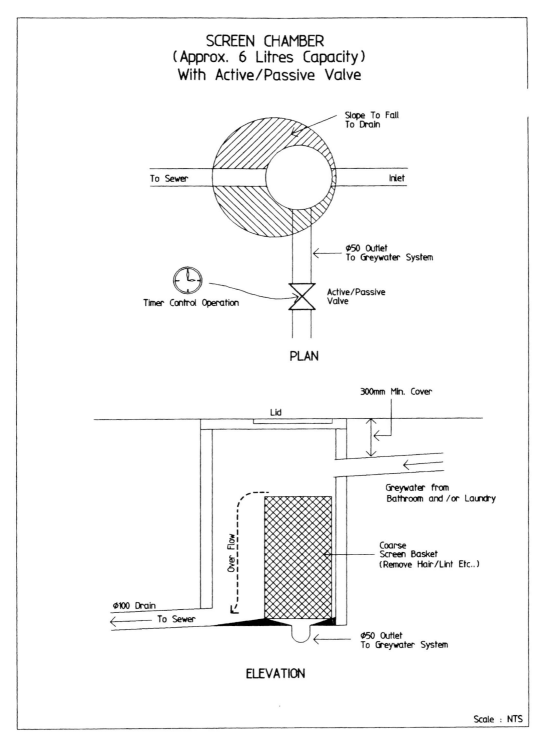

Figure 6.18a: Graphic depiction of a reuse system.
Source: *Model Guidelines for Domestic Grey Water Reuse for Australia*, Research Report No. 107, p. 7.
Reproduced with permission from Water Services Association of Australia.

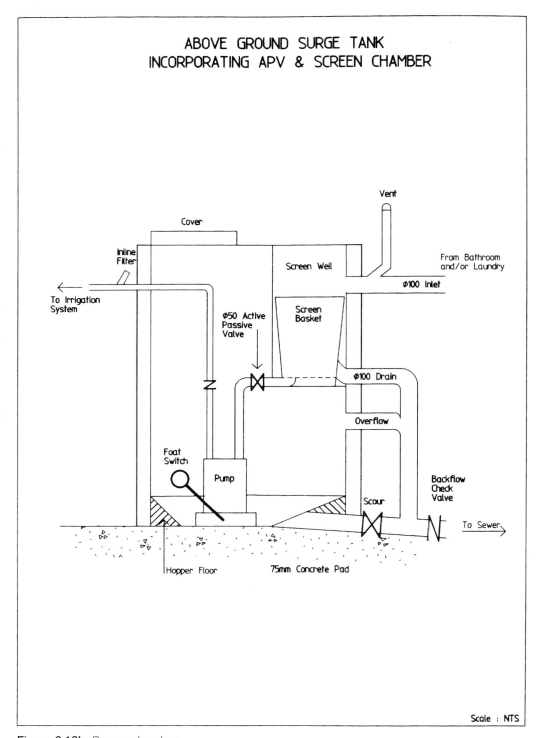

Figure 6.18b: Screen chamber.
Source: *Model Guidelines for Domestic Grey Water Reuse for Australia*, Research Report No. 107, p. 23.
Reproduced with permission from Water Services Association of Australia.

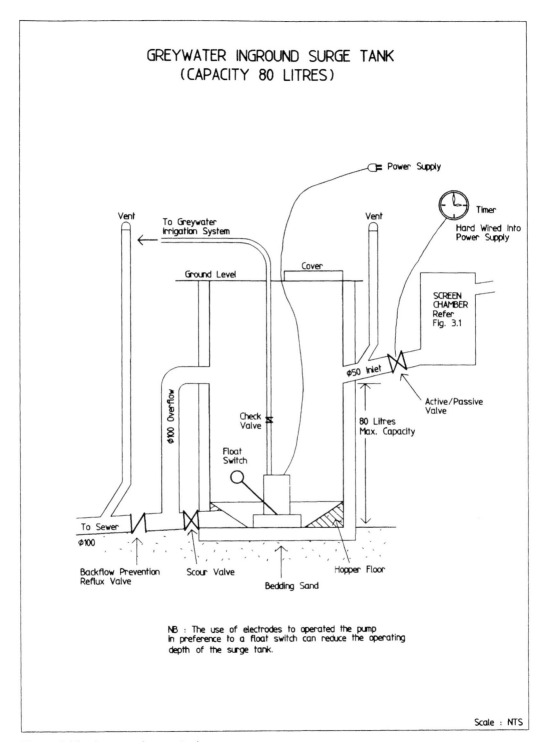

Figure 6.18c: In-ground surge tank.
Source: *Model Guidelines for Domestic Grey Water Reuse for Australia*, Research Report No. 107, p. 22. Reproduced with permission from Water Services Association of Australia.

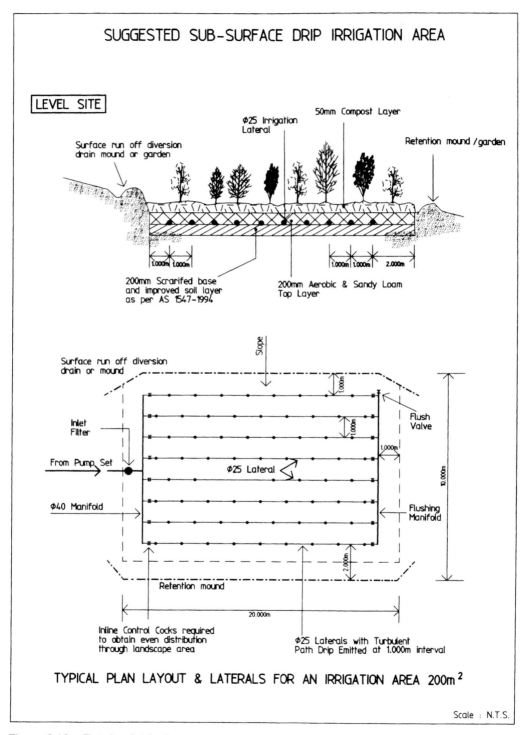

Figure 6.19a: Flat site distribution.
Source: *Model Guidelines for Domestic Grey Water Reuse for Australia*, Research Report No. 107, p. 29. Reproduced with permission from Water Services Association of Australia.

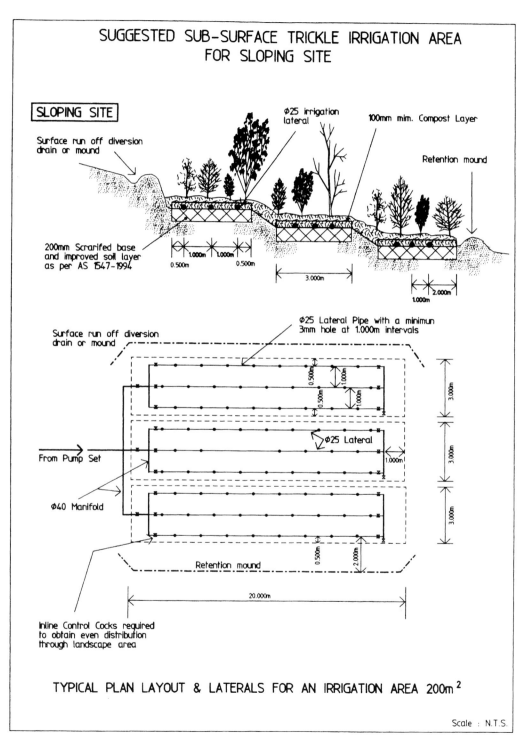

Figure 6.19b: Sloping site distribution.
Source: *Model Guidelines for Domestic Grey Water Reuse for Australia*, Research Report No. 107, p. 28.
Reproduced with permission from Water Services Association of Australia.

Use of treated grey water on gardens

Use treated grey water on the garden only during dry weather because after rain the water can be transported onto neighbouring land. For the same reason, do not overwater using treated grey or black water systems. Remember that you may be legally liable for any contamination of adjoining land. You should also avoid using additional fertilisers when applying grey water or treated black water.

The model guidelines lay down strict conditions for the garden reuse area and your garden distribution system will need to be below ground.

Sample distribution designs suggested by the guidelines
These are simply guidelines. They may not be law in your area. Each state and local authority may have different regulations.

There are several recycling systems on the market. They can be fairly expensive and the associated regulations can be complex, so it pays to do a little research. You will also have to know your soil type. You should first call your local water and sewerage authority to obtain a list of approved systems and any advice on systems suitable to your area. Then you should obtain technical details and prices from the manufacturer. Later, you can get an installation price from your local supplier. It will be compulsory to use a licensed plumber to install any system and associated alterations to your existing plumbing. Ensure that you have an effective system for diversion into the mains sewer during wet seasons or emergencies. If you think these problems are a lot of trouble, you are right. Only the most enthusiastic water-saver will install such a system.

Grey water reuse conclusions

Research indicates that grey water reuse involves the following issues:

- It could save 30–50% of current household usage.
- It would reduce sewage flow loads and environmental problems.
- System installation costs would not be recovered within 10 years, thus making them hard to market.
- Garden distribution systems are costly, disruptive to install in existing gardens and hard to maintain, and in many cases could cause groundwater pollution.
- Hand-basin/toilets are the most viable option, but manufacturers have yet to adapt the technology to Australian conditions.
- Regulatory authorities in Australia are some way from permitting reuse water for flushing toilets and surface garden use as practised in other countries.
- Changes in the situation are likely to be based upon change in household behaviour in the use of chemicals and the technology to deal with faecal matter.
- New technology will become more readily available as water prices rise.
- Disinfection for safe human use can cause soil problems.
- Storage tanks currently lead to increases in micro-organism growth and are thus not recommended even though they would increase the capacity of reuse systems.

- Official research concludes the chemicals are safe to use on the garden except in the case of food crops.

Despite the last point, chemical overload can easily happen in a single application or with repeated applications.

The general acceptance and practice of household water reuse can be driven by regulations, environmental consciousness and/or economics. Water is still cheap and the regulations too rigid. In the words of one researcher, 'If we set the regulatory goals too high, domestic water reuse will fail.'

For example, under health regulations, drinking water may contain up to 150 faecal coliform per 100 mL, but regulations for recycled water set the limit for use on gardens at less than 10 faecal coliform per 100 mL.

Recommended distribution of grey water is below 200 mm from the soil surface, but the soil microbes that can cleanse the water occupy the top 100 mm aerobic layer of soil. Distribution below ground is subject to moisture overload and problems of salinity, as well as clogging of outlets, and problems of water not reaching all plants. The model systems shown on pages 46 and 47 were obviously not designed by a gardener! Costs can be high – pressure pumps are likely to break down after 18 months, and back pressure valves cost around $500 and must be installed by a licensed plumber, who must also check the system annually.

Countries such as Israel, where water costs are high, have more relaxed regulatory restrictions. At present, domestic water around Australia generally costs less than $1/kL. People in the industry estimate that not until costs rise to $2.50/kL will cost-efficient technology emerge, such as in Sweden where dry toilets are in common use. At $2.50/kL users will be more economically driven and reuse practices more widely adopted.

A sample of one set of state regulations is given in Appendix 6. Regulations vary in each state.

Garden watering methods

How and when you water makes quite a difference to the health of your plants and your water budget.

Moisture testing

Use a moisture meter or the one-finger test to see if the soil is moist or damp. Drive the probe or finger well into the soil. If you are not certain whether the moisture goes all the way to the roots, dig a temporary hole alongside the plant and observe the moisture levels in the soil horizon. You can use the same hole to observe how far the water penetrates when you surface-water the plant. Water slowly and see from which soil horizon the water enters the hole. Fill the hole in when finished.

Figure 6.20: The hole was filled with water about 30 minutes before the photo was taken, indicating that both supply and drainage is good.

Frost

If your area is prone to frost, night watering can reduce the impact of lighter frosts by increasing humidity around the plant. This strategy does not work against severe frosts.

Watering frosted plant foliage just before the morning sun strikes the plant can reduce the sudden temperature change that burns leaf tips. The temperature transition due to watering is gentler than the abrupt solar-induced change, which cooks the leaf cells.

Heat stress

In dry weather, observe plant foliage for signs of wilting. Water the quantity you have observed necessary to moisten the root zone. This should allow the plant to recover. It is best, for the plant's sake, to avoid stressing the plant for extended periods, because it can encourage diseases or pest attack.

The bucket brigade can be quite useful as long as you don't have to cart water too far. Two buckets are easier to carry than one. Make sure the plant gets a good soak. Light watering defeats the purpose because it encourages surface rooting, which can dry out easily.

Try the bottle or pipe method to avoid scouring by bucket-watering. Dig a hole alongside your favourite shrub (don't do this for shallow-rooting plants such as azaleas, rhododendrons or citrus). Make the hole big enough for an upturned plastic bottle or an 8+ cm diameter pipe. Fill the bottom of the hole with gravel. Cut the base off the bottle and place it upside down on the gravel. Fill it by bucket or hose.

Foliage or root watering

It is far better to water the soil, so that the plant roots get water, than to water the foliage directly. Remember that most plants have their own means of transporting water to leaves from their roots. This rule applies particularly on hot days when the sun is likely to shine on the wet leaves, overheat them and cook them (see Appendix 1).

The rule also applies to recycled water, which may contain chemicals. Chemicals can accumulate in the soil, doing damage in the long run.

Lawns

Established lawns

Lawns can be trained to a cycle of watering during summer. The best method is to:

- watch the lawn for signs of stress wilting. It will stay alive for a day after the first signs
- water for enough time to wet the soil to a depth of 15 cm. Test with a probe or make a test dig to ensure moisture has penetrated to that depth
- observe the lawn until the next stress time. You will then know how long to space your watering times. It could be 4–5 days between watering. The times will vary depending upon factors such as soil type and depth, lawn species, the weather pattern and the efficiency of your sprinklers. Make sure you set your sprinkler timer to water at night
- renew the organic matter every spring by coring and applying a topdressing of worm castings or blood and bone with lime. Rake in as well as you can
- mow your summer lawn high enough that only the top third of the grass blades is cut. This will stimulate new root growth and provide shade insulation from the remaining grass blades.

New lawns

There is a new range of dry-tolerant turf grass species. The species vary for each climate zone. Read the packet carefully, and remember that many of the creeping lawns will need control at the edges to keep them out of garden beds. Your local agriculture, primary industry department or water supply authority can advise you on the new varieties best suited to your area.

New lawns consisting of turf rolls are initially water-efficient compared to growing from seed, but you should ensure that there is a good depth of friable soil rich in organic matter to provide a foundation for healthy growth. Do not compress the foundation so much that it becomes an impenetrable barrier.

Water budget

For an overall household budget see the water audit in Appendix 7.

Create a garden water budget based on your water inventory and try to stick to it.

Fruit trees in fruit require 400–600 L per week. You can reduce this after harvest but do not be miserly, because water during autumn is important for the tree's health next spring. Large deciduous trees may use more water through transpiration. The best approach is to reduce their allowance by spacing the watering period anywhere from one week up to one month in severe drought, while keeping a sharp eye out for stress signs such as leaf wilt or yellowing (as if autumn were approaching).

Count the number of large trees and shrubs and multiply them by 400 L to get your weekly water budget. All plants vary in their need for water. Tomatoes will use up to 2 L per day with fruit on. Oaks may use up to 100 L a day. Eucalypts and wattles are among the most economical of trees, with mallees being the best eucalypt. You should not have to water your mallees at all. If they die, most other plants will have preceded them.

Watering times

Water mainly towards or just after dusk to allow the water to penetrate the soil around plant roots with minimum loss from evaporation. Of course, some of your bucket watering may have to be done during the daytime, as buckets fill. Common sense should prevail and it is best to bucket-water plants in shade. Water into 'ground bottles' or through deep mulch. Develop a rotation system so that each plant gets a deep periodic watering, not a light frequent watering which encourages surface rooting.

You may water for a few minutes on frosty mornings to reduce frost damage.

Automatic sprinklers can be timed to water at night. Start the watering cycle around dusk so most of the water penetrates the soil well before sunrise. If you have a power failure overnight you will have to reset the timer.

Light rain does not provide enough water for most plants. It will freshen them up temporarily, but it may not penetrate the mulch. Do not turn off your timer for light rain, but if the rain is reasonably heavy you can turn it off.

Chapter 7

The theory and practice of mulching

Nature has been mulching for millions of years in a variety of ways, influenced by climate, soil type and plant species. Plants drop leaves, animals die or excrete on the soil surface to form mulch as part of nature's recycling system.

Yes, mulch is nature's cast-off. Earthworms, insects, fungi and microbes convert those cast-offs into soil nutrients that make up the plants' menu. Leaf litter in forests and grassland forms a living blanket over the soil, protecting and nurturing the plants that grow there. So, if you wonder what mulch is and what it does, the best answer is to study nature.

In the home garden mulch plays many roles.

- It insulates the soil, plant roots and small creatures from temperature extremes, maintains a healthy level of soil moisture and reduces the need to water the garden. This is a true micro-climate – the plant comfort zone.
- It provides a habitat for a myriad of small life forms, which in turn busily convert it into soil and plant nutrients.
- It protects soil from water and wind erosion.
- It suppresses most weeds that might compete with the plants you wish to grow and thus saves you time in the garden. Or it can be used to smother existing weeds so that they become a green manure crop.
- It looks better than bare soil.
- It offers an opportunity to emulate nature and recycle many waste products.

How deep should it be?

The depth of mulch required depends on the soil, plants, season and background climate.

Sandy soil will need more mulch than rich organic loam. Deep-rooting natives will need less mulch than shallow-rooting exotics. In summer, mulch should be a lot thicker than in winter or spring. Tropical gardens require far less mulch than those in climates with long dry summers. But mulch will disappear more rapidly in the tropics and subtropics due to a more active conversion by earthworms and their co-workers.

Vegetables need mulch from their seedling stage to maturity, but in early spring mulch should be kept thin until the soil has warmed up adequately for healthy deep root growth.

Mulch depth also depends on the type of mulch material. Coarser material will need to be deeper to stop light and weed penetration.

'Thin' mulch is about 50 mm deep while 'deep mulches' can be up to 200 mm thick.

The depth you choose will be a matter of judgement based on advice and your experience.

Insulation

You will see later (p. 64) how plants use natural self-mulching as a water conservation strategy. The theory behind insulation is not new – for example, we use insulation in our homes to moderate temperature extremes. It is the air within the open-textured

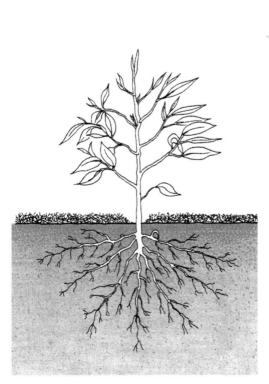

Figure 7.1a: Thick mulch protects the soil from wind and sun, retaining moisture around the roots. The plant is healthy.

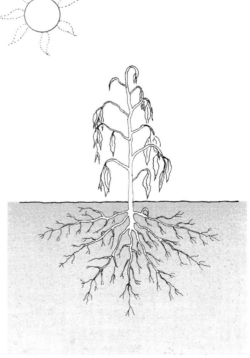

Figure 7.1b: Where there is no mulch, sun and wind action reduce soil moisture. The soil dries and the plant is under stress.

material that helps provide the insulation. The same happens with garden mulch. The enclosed air combined with the mulch material reduces conducted, convected and radiated heat transfer.

In other words, the mulch material of leaves, straw, fibre, grain or pebbles provides a thermal blanket for the soil, which remains at a fairly constant temperature and moisture level. If the soil is at field capacity (see Appendix 1) when mulched it will retain that condition for much longer than bare soil. In general, the thicker the mulch the longer the soil will hold its moisture.

Mulch as habitat and food

'Earthworms are the most important animal to man' CHARLES DARWIN.

Earthworms perform their important soil enrichment function under mulch.

The moderate temperature and excellent moisture retention of soil under mulch favours the development of earthworm populations.

If mulch also contains nutrients for the earthworms they will reward us by spreading valuable worm castings throughout the soil horizons.

From a gardener's point of view, there are two types of worms – earthworms and compost worms. Earthworms burrow into the soil depths, coming to the surface periodically to feed on mulch and redistribute its goodness when they return to the depths. You cannot reintroduce earthworms into the soil, as they will die. The good news is that there are always many present in the soil in cocoons, waiting to hatch when nature or the conscientious gardener provides the right moisture and temperature conditions. Mulch helps to retain those conditions and extend the life and beneficial activity of earthworms.

Compost worms have evolved in the leaf litter of forests and grassland. They like a nutrient-rich situation provided by such materials as manure. They do not survive in soil alone. You can purchase them from worm farmers or, if you are patient, they will migrate over a few weeks to any part of the garden that you water, mulch and manure well and will breed there of their own accord.

The insulating mulch with its natural fertilisers will help the worms survive and work for you a lot longer. Soils with worms have up to 20% better water retention, and they love the micro-climate that good mulch provides. Every time you mulch, the earthworms will increase the soil depth and organic matter, without the addition of artificial fertilisers.

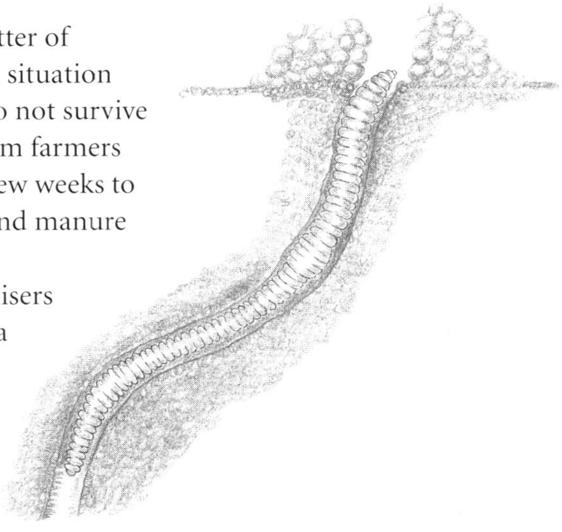

Figure 7.2a: Mulch as habitat and food.

In dry times many worms die. Some species, which inhabit surface layers, will perish. Others, which burrow more deeply, will hibernate until times are better. The micro-organisms also go into decline awaiting moister conditions.

Worm bank

Even the worms that die leave a survival heritage of eggs to hatch when the environment signals conditions are right. If your soil has dried out during drought do not begin watering if you cannot maintain the rate of water, because the eggs may hatch and the future worm generations will perish upon the return of dry conditions. You must conserve your worm bank.

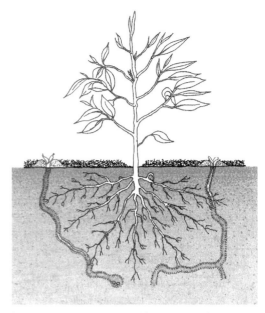

Figure 7.2b: Worms working near a plant root system operate in a symbiotic way. They prune any dead roots and build soil structure and nutrition.

Worms in soil without mulch

The worm population in soil without mulch will fluctuate with seasonal conditions. The population will increase in spring, die off in summer, recover in autumn and hibernate or perish in winter. The population will never be as great as it would with mulch. The soil will have fewer worm burrows and less organic matter, and will not be able to store the water needed for healthy plant growth.

Worms in soil with mulch

The worm population under mulch remains more even throughout seasonal changes, although it must be understood that, with the exception of the tropics, worm populations do not thrive in winter.

The soil under properly managed mulch becomes heavily populated with earthworms that burrow very actively, feeding on the mulch and transporting the organic matter to the soil depths. The additional burrows and increased organic matter helps the soil retain significantly more water than bare soil, and it will keep getting deeper the more you mulch.

Soil protection

Mulch is an ideal means of preventing your garden from looking like a miniature Grand Canyon or Sahara Desert. Appropriate mulch can reduce water erosion to zero in sloping ground which is prone to water erosion after heavy rains. It can also prevent soil loss due to strong winds, especially in sandy soils.

Weed suppression

Correct mulching stops the sun from penetrating to any seeds on the soil, thus reducing the germination rate of weeds. If you take the trouble to remove all weed seed sources, take out taproot weeds and renew your mulch at least annually, you should be able to reduce weed growth to manageable levels. By reducing weeds you reduce the competition for the moisture and nutrients that you hope to preserve for your favoured plants.

Weeds as green manure

If weeds exist before mulch is applied, you can treat them as a form of green manure. Most green manure crops are dug in, but I prefer to flatten the weeds, then place manure and a few handfuls of lime, blood and bone, and worm castings on top. Water well, then cover with newspaper at least three pages thick and overlapping, and finally apply your mulch about 10 cm thick. This can be planted into after three months or so. It may sound like a lot of trouble but it beats digging out weeds! Also, what is little recognised, even by the experts, is that weeds have a wide range of mineral content. The act of smothering the weeds instead of digging the tops in allows the roots, which are the richest part of the plant, to break down in the soil, thus feeding the worms and microbes to create beneficial plant food and deepen the soil. By the way, you can also plant your own green manure crop and treat it the same way. However, remove any weeds with taproots such as dock before flattening the tops.

This technique is further explained in my book *Green home recycling*.

Mulch as a cosmetic ground cover

By forming a textured natural surface, mulch can provide an attractive groundcover and backdrop for your flowers and shrubs. It makes the garden look tidy and natural at the same time. You can create special texture and colour effects by choosing from the wide range of mulch materials.

Figure 7.3: The mulch of washed pebbles emulates a dry streambed. In this well-designed garden at the Basaltica demonstration site in Altona, a drain flows under the pebbles. **Source:** Basaltica

Applying mulch

Figure 7.4: The mulch is thick around the chilli seedlings but not around the base of the citrus, to avoid collar rot.

'Mulch that is thin cannot win' – it must be applied with ample thickness. Several factors are important for successful mulch application.

- You must have ample quantities for the job at hand.
- Do not mulch up to the base of a plant susceptible to ring rot and other disorders of the bark caused by constant dampness. Such plants include camellias, azaleas, rhododendrons, citrus and deciduous fruit trees. You will probably need to weed around the base of susceptible plants. Some plants are not affected by mulch around their base. For instance, tomatoes and potatoes like mulch up to the stem, and in fact potatoes need it.
- Make sure the soil is really moist before you mulch. By first giving the soil a good soaking you will start activities of soil life in the same way as a normal spring.
- Every year, vary the mulch ingredients that you apply in one location to avoid an overload of particular nutrients. Exceptions are compost or multiple blends.
- With the exception of natives such as banksias and grevilleas, make sure that the soil has plenty of organic matter to help the microbes and earthworms work, and that it can hold the maximum of moisture. Ideally, it should not be compacted so that the plant roots can penetrate the soil. This particularly applies to vegetables. If your vegetable plot includes paths and beds up to 1.2 m wide, you will not need to stand on the soil to weed or harvest.
- Make sure the soil has warmed up in spring before you mulch. You should use a soil thermometer to measure soil temperature. If you need to know the best temperature for individual flower or vegetable seeds look in the packet direc-

tions or consult a good garden book. Mulch applied over cold ground is liable to keep the ground cooler for much longer because of the insulating qualities of the mulch.
- To overcome this problem, either apply mulch after the ground has warmed up, or bury leaves and other compostables 300 mm (12 in.) deep in the ground to warm the soil by compost action. Once the plant is growing you can cheat and increase the thickness of the mulch. The trick is to mulch thinly at first then, once the plant is growing vigorously, increase the mulch depth.
- Soil aeration will help soil moisture retention and the penetration of fertilisers. You should lift the soil with a fork without turning the soil over. This will cause minimal disturbance to worms and other soil life. You can incorporate manure or compost at the same time. Use gypsum in clay conditions to encourage soil particles to form granules and improve soil structure and aeration.
- Make a layer of manure or compost on the soil surface. Incorporate some or all by fluffing it into the top layer with a fork. This helps hold precious gases within the soil.

Place a layer of newspaper over the top of the manure or compost to discourage weed germination and protect nutrients, then mulch over the paper. Do not manure in dry times or when plants cannot be irrigated. Manures and fertilisers promote the type of growth that is not drought-hardy, and they can burn roots they come in contact with. To hold the benefits in the soil, apply a deep layer of mulch, such as 15 cm (6 in.) deep or, if you are using newspaper as shown in Figure 7.6b, 10 cm is adequate.

Vegetables and flowers

If you grow any vegetables, continue to grow what you need. Concentrate on vegetables that can be well looked after. Mulching methods for vegetables and annual flowers are similar.

Plant your seeds in the usual manner and water well. Then place a wet bag over the top. Inspect morning and night until the seedlings emerge, then remove the bag. Sprinkle clippings from your lawn lightly around the sprouting seeds and place a frame of the lightest shade-cloth over the top. As the seedlings grow to transplant stage, gradually harden them by removing the frame for longer and longer periods.

Mulch your paths between beds as early as possible. Place newspaper directly over the path, covering any weeds. Use up to six sheets thickness. Wet as you lay if conditions are windy, but it is best to lay newspaper during a calm period. Cover with mulch material.

Figure 7.5: Your reward.

When planting vegetable or flower seedlings you have two mulch options.

1. Water your seedlings well, then apply any appropriate fertiliser in a band away from the roots so that they grow toward it. Hoe it in. Apply your mulch cover over a layer of newspaper once the seedlings are established. Plant your vegetables as early in the spring season as you can to get the best from any rain that follows.

2. Place the manure or compost on the soil, cover with newspaper, then mulch. When you wish to plant, simply make a hole in the mulch and paper where each seedling is to go and water it in well. You should not have to water for a few days unless there are signs of heat stress.

Figure 7.6a: A damp bag laid over the seed bed assists germination and saves watering. Remove immediately seedlings appear and water regularly.

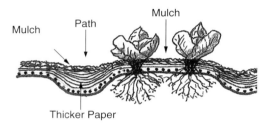

Figure 7.6b: Mulch extends over beds and paths.

Figure 7.6c: Mulching not yet completed.

The photograph in Figure 7.6c shows the newspaper laid next to the row of seedlings with the mulch being applied on top of the paper. Later the mulch will be applied thicker between the rows and between the seedlings when they are more robust.

The theory and practice of mulching | 61

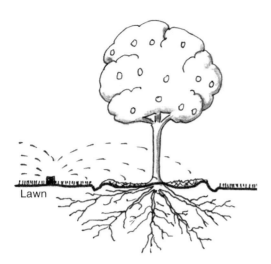

Figure 7.7a: A specimen fruit tree placed in the lawn gets the benefit of lawn watering.

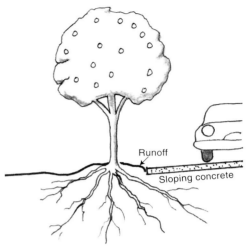

Figure 7.7b: The fruit tree can benefit from the runoff of water from the driveway.

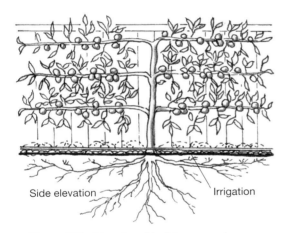

Figure 7.7c: Espaliered fruit trees can be watered individually by mini-sprinklers placed along the length of the trees.

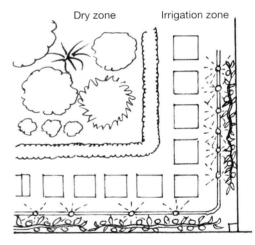

Figure 7.7d: The plan shows how the espaliered fruit tree can be watered as part of an elongated wet zone alongside a dry-zone planting. The pavers form a border between the dry and wet zones.

Frost protection is important for most vegetable crops and some flowers. Mulch areas suffer from frost more than the rest of the garden. You can overcome this problem by covering plants with a cardboard box overnight. Gradually harden the plants by lifting the sides one by one then, after a week or two, remove the box altogether. But watch out for frost.

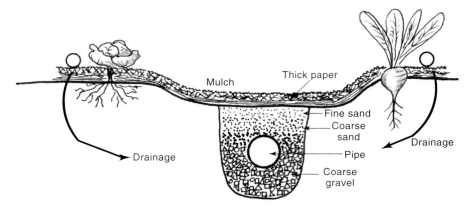

Figure 7.7e: Drains can be placed under paths to collect and recycle the nutrient-rich water (see pages 100 – 102).

Trees and shrubs

Trees and shrubs in general can tolerate coarser mulch materials. They resist most, but not all, pests that lurk in the coarse mulch depths. Apple and pear pests spend part of their life-cycle in the soil under the tree so you would not choose coarse mulch for those species. The main thing to consider is not to build up the mulch around the base of the tree or shrub to avoid ring rot of the bark.

Non-organic mulches

This far we have been considering organic mulches that eventually become incorporated into the soil they cover through the action of microbes, earthworms and other small creatures. Non-organic materials such as gravel do not do this. They have special uses and shortcomings.

Figure 7.8: Washed pebbles or rocks have their place and are useful in trafficked areas.

Gravel, large pebbles or washed stones are useful when mulching around plants that require little nutrient input or even disdain fertiliser altogether. Many natives or desert plants grow well in low-nutrient situations similar to that in which they evolved. The mulch in a dry zone which is low in nutrients provides a cosmetic effect and erosion control.

The gravel mulch works because it helps the soil retain moisture and evens out temperature extremes. It also allows most moisture that falls on the mulch surface to drain through to the soil below.

Gravel or stones can also be used in wet zones, but if you wish to add nutrients a ground cover of weedmat under the gravel will be needed to suppress any unwanted growth.

Figure 7.9: Pebble gaps offer a habitat to weeds, earwigs, slugs and other pests.

Shortcomings of gravel or stone mulch
Long experience has shown that the beautiful pebbled mulch works well where little or no rain falls but can create problems where rain falls. Dust and seeds from weeds lodge between the stones over time. When it rains the seeds germinate in this perfect seedbed. The stones then become an impediment to weeding. Under these circumstances even weedkiller becomes a short-term solution. The only realistic solution is a costly complete renewal. For this reason, pebbles and larger stones or gravel are mainly limited to dry areas.

Living mulches

Living mulches present another option for the waterwise gardener. Properly applied they can be a cheap, labour-saving and near-permanent solution to many landscape problems. You must thoroughly understand the purpose of the mulch for each situation.

- Soil type is very relevant for some plants, whereas other plants will do well in any soil. Soil pH is important. Most native plants will do well in neutral to moderately acid soil. Many herbs like neutral to moderately alkaline. But there are exceptions to both rules. Some plants prefer light soil, while others prefer heavy soil or are happy in any texture soil. Most prefer good drainage. They come from many places around the world – mountainous areas, coastal cliffs, dunes or deserts.

- Aspect is moderately important, but full sun or shade is more important in determining whether a living mulch will do well.
- Some living mulches are used to prevent soil erosion.
- Others are mainly cosmetic.
- A mulch does not have to be a low creeper, although many are. Some shrubs spread and cover so well that nothing will grow under them. You might call them a tall mulch.
- There are even mulches for kitchen gardens. Legumes make ideal mulches for food plants because of their capacity to nodulise nitrogen on their roots.

You will find most of the living mulches listed under groundcovers in the appendices.

Figure 7.10: Clover makes an excellent living mulch at the base of fruit trees.

Self-mulching plants

Most plants are self-mulching to a degree, but we are interested in the selfish species that are more effective in suppressing competing weeds and retaining moisture for their own use.

Plants such as pines drop needles that form dense long-lasting carpets and are among the most effective self-mulchers. Actually, there is generally a dual mulching action taking place in a pine forest. First, the dense tree canopy reduces the penetration of light and winds so that weeds seldom grow and evapotranspiration at ground level is reduced considerably. Second, this is aided by the superb insulating qualities of the deep long-lived needle litter.

Some gardeners collect pine needles from plantations to mulch their home gardens. You should keep in mind that needle mulch can eventually turn soil acid, so use this mulch only on acid lovers, or modify its pH by adding lime or blending the needles with several other mulch ingredients.

Other good self-mulchers include:

- some acacias, which drop many leaves and seedpods that all enrich the soil
- allocasuarinas and casuarinas, which also drop many needles
- many deciduous trees, such as oaks, which have a multiplicity of large leaves that form a soil-building carpet every autumn.

Figure 7.11: Allocasuarina needles.

Mulch materials

The list of mulch materials that follows is not as comprehensive as those in other sources, which include some nutrient-rich materials that, for two reasons, I consider are more valuable when applied directly to the soil beneath a protective mulch.

1. Nutrient-rich materials exposed to direct sunlight, wind and rain soon lose their nutrient value.
2. Such materials provide a good bed for weed seeds, unless more inert materials cover them. Manures may also contain viable weed seeds that will germinate under the right conditions.

Garden and household collectables

There are many mulching resources in your garden and house. Prunings from fruit trees, dead tops of perennials, hedge clippings, old corn stalks and any cuttings from shrubs and trees can be collected. Avoid diseased material and rose cuttings, which can spread disease further. Ensure that clippings of such trees as hawthorn are dried out so they cannot grow in unwanted places.

Newspaper, cardboard, carpet, carpet underlay and other old natural materials make great mulch underlays.

Mulcher

The ideal method of breaking down the coarser material is to shred it through your mulcher then spread it out to dry. Make sure you wear safety goggles.

Lawn clippings

Lawn clippings are useful for some mulch activities once they have dried out. Treat clippings, as a farmer would hay. Leave the catcher off the mower and leave the clippings for a few days then mow again with the catcher on. Try to avoid compression when storing clippings, and keep them dry. Lawn clippings serve your drought strategies best if kept fluffy. Soggy clippings will not last. Lawn clippings have limited value on their own as they blow away easily, so blend them with other coarser materials.

The green clippings can be used to top up shallow mulches, but because they tend to draw nitrogen out of anything they are placed upon it is best to blend in a small amount of nitrogenous material such as aged poultry manure.

Ensure that any lawn clippings do not contain grass seed.

Outside collectables or purchases

'One person's waste is another person's resource' is an old saying that still holds, even in today's more recycling-conscious society. The reasons are simple. Some people are not as aware of reusing waste products as they should be. Some wastes are so plentiful that the producers don't know what to do with them. Some wastes are too far from recycling industries for it to be economical to transport them. Recycling costs also fluctuate, causing gluts and shortages.

There is a strong likelihood that you can find a suitable industrial waste that can be used as mulch. Stable manure is one that is readily available in many areas.

When you purchase mulch you are entitled to a clean product that is weed-free. Unfortunately this is not always the case. You can go to your local nursery and see clean mulch in the yard there, but perhaps the bulk supplier's yard may be next to a paddock of rampant weeds at seed stage. You will never know – until these weeds start to sprout in your mulch.

The solutions are:

- to purchase from reputable dealers
- to make a point of asking if the material is weed-free
- to purchase just after winter, before any potential weeds go to seed near the bulk supplier's stockpile
- to obtain your material just off the production line.

Collecting mulch

Some materials are best bought in bulk, if you can afford it. Buying in bulk works out cheaper per unit of volume. Truck delivery is an option if you can afford it and you do not wish to cart material in a trailer. You can get truckloads as small as 4 m^3 or large as 10 m^3. Make sure there is access to your garden and you have time to distribute the mulch, or room for storage.

Using the trailer instead of having mulch delivered by truck is a false economy, as I have found to my cost. I used to cart by trailer all the time, until one day the load was too much for my transmission. The cost of repair? Many times the cost of delivery. Of course many mulches are lightweight and put no strain on your vehicle. I still cart stable manure and recycled straw by trailer. When carting such material ensure that it is covered to avoid it blowing about. You not only face losing material but you may also incur a traffic fine.

Wastes

Check your area for industrial wastes, but be careful to avoid pollutants and know the qualities of the product first. For instance, coffee grounds are highly acidic. Quarry dust can be of great value. In the tropics, sugarcane waste is in plentiful supply. On one occasion I used poppy waste as an excellent nutritious mulch, only to find poppy flowers emerging. Needless to say I quickly destroyed them.

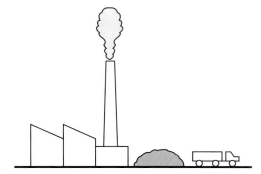

Figure 7.12: Many industrial wastes have good mulch potential.

Warning

Be extra careful when handling industrial or other mulch materials that may give off fine dust that can cause serious diseases such as Legionnaires' disease. It is best to wear a filter mask when handling mulch or compost.

Mulch as habitat for pests

Mulch by its very nature provides a great habitat. The open structure allows many small creatures to make their homes in the warm moist dark depths. A short time after mulch has been spread the inhabitants move in and others come as co-inhabitants or predators. Before long a microcosmic world begins a complex existence that would not take place on bare soil. Microbes, fungi, slaters, earthworms, centipedes, spiders and many more tenants seem to come from nowhere.

Some of these guests, such as snails and slugs, are the enemy of the gardener. Some, such as earthworms and centipedes, should be welcome. Some are hard to identify as enemy or ally.

There are several strategies that can be adopted.

- All sensitive seedlings should be well established before snail-favouring mulch is spread around them. The more mature a plant is the less attractive it is to snails and slugs.
- Careful targeted baiting in the spirit of integrated pest management should be used.
- Welcome birds even though they make a minor mess of the tidy mulch.
- Select mulch such as sawdust that is not quite as open as, say, straw.
- Set up a defence perimeter by laying a trail of limil (slaked lime, hydrated lime or bricklayer's lime) around the bed. Many pests do not like to cross it. Avoid breathing its fine dust. It kills snails and slugs (but also kills many helpful creatures) by drawing moisture from their bodies.

Mulch materials list

Bagasse (sugarcane waste)
Cane waste has good insulating qualities, but be careful handling it. The dust of bagasse can be harmful to lungs, so wear a face mask.

Its advantages are that:

- because of its texture bagasse is less liable to harbour pests such as snails
- it will break down over a year to make a valuable addition to soil organic matter
- it is weed-free.

Figure 7.13: Bagasse heaped at a sugar refinery.

Its disadvantages are that:

- it is flammable
- it must be kept 10 cm away from the base of plants.

Mix bagasse with other materials for best results. Try a little aged animal manure in the blend. As Figure 7.13 shows, there is plenty of it in sugarcane growing areas.

Bracken
Bracken is a natural mulch material that can be mulched down to a relatively fine size for general use. The fine bracken does not last as long as the bracken of stalks slashed from the paddock where it grows. Slashing temporarily removes an agricultural nuisance that can impose problems for grazing stock, so using it solves two problems at once. I do not suggest that you gather it from natural bushland.

Advantages are that:

- it is a good insulator, used coarse among larger plants or fine among smaller specimens
- it is rich in potash, phosphorus, manganese, iron, copper and cobalt, so it should help fruiting plants.

Its disadvantage is that:

- it is very volatile so it should not be applied in high fire-risk situations.

Carpet and natural carpet underlay

Nothing covers like carpet, but make sure it is natural material if you are using it in food-growing areas. Artificial fabric can be used in pathways and shrub or tree areas. It is best to cover the carpet with a cosmetic mulch.

Its advantage is that:

- it will smother just about anything including blackberry seedlings and couch. Make sure cover is complete so that runners do not emerge from the sides.

Disadvantages are that:

- carpet may look good on your floor, but it looks completely out of place on garden soil. Hence the advice to cover it with a better-looking mulch
- once laid, carpet can be a tough barrier to cultivation or planting through mulch.

Chipwood

Chipwood is good for most home shrub beds.

Figure 7.14: Chipwood can even be made from old fence posts.

Advantages are that:

- it provides a valuable cosmetic appearance
- it will not blow about and if placed over weedmat it will keep weeds down
- you can water through chipwood but try to avoid wasteful sprinklers. An underlay of manure will help plant nutrition
- it will break down over a long time to make a valuable addition to soil organic matter.

Disadvantages are that:

- it is very volatile so it should not be applied in high fire-risk situations
- it can harbour snails and slugs.

Coffee grounds
Coffee grounds are best mixed with other materials.
   Advantages are that:

- it is weed-free
- it has a rich brown appearance
- it will break down over a year to make a valuable addition to soil organic matter.

Its disadvantage is that:

- it is highly acid so it should be mixed with lime or wood ashes, as well as other materials.

Compost
Compost is an extremely valuable mulch because it breaks down to enrich the soil.
   Its advantage is that:

- it has a wide range of nutrients.

Disadvantages are that:

- compost is degraded by direct sunlight and leached by rain
- it takes a lot of compost to mulch
- any weed seeds present will germinate and defeat the purpose of smothering weeds.

Figure 7.15: One of the most efficient and sturdy compost tumblers on the market.

The soundest way to use your compost is to water it well and cover it with another less valuable mulch. Then let the worms do their job for you.

Cow manure
Slightly aged cow manure is a great soil conditioner, providing good plant nutrition. In particular, it is a great source of nitrogen, phosphate and potash. The solids comprise complex carbohydrates such as cellulose and lignin. If allowed to break down on the surface of mulch, they release carbon dioxide and hydrogen into the atmosphere, but if they are applied as a layer under a more inert mulch this loss is reduced.
   Manure of all sorts is a beneficial addition to all compost. Cow manure is one of the gentlest manures because it has been digested in the cow's several stomachs.

Eucalyptus mulch (eucy mulch)
Eucalyptus mulch is a good choice for native gardens, but avoid using it around exotics. Its open ventilating texture requires fairly dense layering or weedmat underlay.

Figure 7.16: Eucalyptus mulch.

Advantages are that:

- it has a valuable cosmetic appearance
- it is an economical cover for low-maintenance areas.

Disadvantages are that:

- it can release tannin
- it is a definite fire risk.

Feathers

Feathers make an excellent constituent of compost, but should not be used alone as a mulch.

Advantages are that:

- they are very high in nitrogen (over 15%)
- they can be obtained from poultry processors at low or no cost
- they are long-lasting.

Disadvantages are that:

- if sourced from processors they will be contaminated with meat, which will attract rodents and flies
- they are lightweight and can blow about the garden
- they look attractive on birds but not on the ground.

Feathers are best processed in a shredder and composted in a rodent-proof barrel composter before applying under a more attractive mulch cover.

Grass clippings

Lawn clippings are the most plentiful mulch resource in most home gardens.

Advantages are that:

- clippings are in continuous supply
- they are high in nutrients, particularly if the lawn has lots of clover
- they are easy to handle.

Disadvantages are that:

- because of their water content they turn to a slippery mess if used green on their own
- fresh clippings draw nitrogen from the soil
- they break down quickly.

Figure 7.17: Grass and other mulch materials used to advantage in Jade Forrest's Garden of Life at Eudlo on the Sunshine Coast hinterland.

It is best to use them as only one part of compost, or use them over paper or cardboard in areas where there is no foot traffic. They can also be used to thicken mulch by layering them under paper and a more inert material.

If dried, they will last longer in an area which does not get overhead watering.

Hair (human or animal)
While hair is a good ingredient in compost, it is not recommended as a mulch on its own.

Hay
Hay is generally purchased in bales after the farmer has mowed pasture, cured it in sunshine and compressed it into bales. There are different types of hay, depending on the crop. Hay usually contains seeds, but when they germinate in the mulch it is little problem to pull them out. Try to see that there are no bad weed seeds such as thistle present. Rain-spoiled hay is cheaper but has lost much of its nutritional value. There are several types of hay.

Figure 7.18: Hay.

Lucerne (alfalfa) hay
Lucerne hay is generally coarser than the other hays.

Advantages are that:

- because it is a legume, lucerne hay is very high in nitrogen and is a great soil improver, especially in the vegetable patch
- earthworms love it
- it is a good soil insulator.

Its disadvantages is that:

Figure 7.19: Lucerne hay.

- its rough open texture makes a good breeding ground for pests such as slugs and snails.

It is best applied after seedlings have become well established. Chopped lucerne is easier to handle but breaks down quickly.

Clover hay
Clover hay is also a legume. You will seldom get clover alone – it usually comes as part of a pasture mix.

Its advantage is that:

- it is a great soil improver and insulator.

Disadvantages are that:

- it provides a habitat for snails and slugs
- it probably contains weed seed.

Lupin hay
Lupin hay has many of the qualities of lucerne and clover hay. It is a good choice and may be cheaper.

Grass hay
Although not as rich in nitrogen as the other hays, it still helps with insulation and nutrition.

Its advantage is that:

- it is not as expensive as the other hays.

Disadvantages are that:

- it is marginally more flammable than the others and suffers the same pest problems
- it probably contains weed seed.

Hessian
Formerly plentiful, hessian has become less available since artificial fibre bags came into general use. It may still be found as product wrapping, and is available in commercial rolls.

Advantages are that:

- hessian effectively smothers weeds
- it is a natural material that eventually breaks down to form soil organic matter.

Its disadvantages are:

- scarcity and cost.

Hessian would need to be covered with a cosmetic mulch.

Lake weed
Some towns clear their lakes of weed. Check to see there are no significant pollutants. Do not apply around food plants.

Its advantage is that:

- it is rich in organic matter and is best used as a mulch underlay.

Disadvantages are that:

- it can attract flies and is better used as a constituent of compost
- it is difficult to determine if pollutants are present.

Leaf litter

Leaf litter makes superb mulch. Deciduous leaves are best, such as oak or apple leaves which are rich in nitrogen, phosphorus and potash. Apply it thickly to a depth of 20 cm.

Advantages are that:

- it is a natural material that breaks down within the year to form rich soil organic matter
- you can apply it on flower, vegetable and other beds.

Its disadvantage is that:

- if applied direct it can go soggy, so it is best to allow it to dry first.

Figure 7.20: Lake weed.

Figure 7.21: Oak leaf litter is one of the best soil enrichers.

Manure

In general, aged manure is best applied under a more inert mulch so that plants get maximum benefit. Manure of all sorts is a beneficial addition to all compost.

Its advantage is that:

- it is rich in nutrients and has all the qualities of good mulch.

Disadvantages are that:

- manure is degraded by direct sunlight and leaching rains
- it takes a lot of manure to mulch
- any weed seeds present will germinate and defeat the purpose of smothering weeds
- it can attract flies.

The quality of manure depends on the source. See also cow, pig, stable manure and poultry litter.

Mushroom compost

This has similar qualities to first-rate compost and manure except it is unlikely to contain weed seeds. Cover with a mulch. Cost may be a consideration.

Paper

Paper makes an ideal sunscreen to eliminate weed seed germination, but it is best covered by a more attractive mulch both to improve its appearance and to stop it blowing about. A thickness of two sheets of paper can smother weeds yet allow the soil to breathe.

Advantages are that:

- newspaper covers quickly
- newspaper is ideal to place over manure and under mulch
- worms love it.

Disadvantages are that:

- it breaks down quickly
- it is hard to apply in windy weather.

Figure 7.22: The mushroom-growing medium is a highly nutritious mulch best used as an underlay covered by a cheaper, more durable material.

Do not use glossy or coloured paper because of the chemicals involved. Shredded paper will still need a cover of another mulch.

Peanut shells

Peanut shells make an excellent ingredient in compost.

Advantages are that:

- they are high in nitrogen and potash, and contain useful levels of phosphorus
- they are an excellent insulator if laid about 15 cm thick
- they should break down over a year
- they are weed-free.

Figure 7.23: Peanut shells make a long-lived mulch.

Disadvantages are that:

- the open texture can provide a habitat for snails and slugs
- some people may not like their appearance
- they are flammable.

The best use is to include them in the compost layer next to the soil and overlay with a finer material.

Pea straw

Pea straw makes a brilliant mulch for more advanced vegetables.

Advantages are that:

- it is high in nitrogen, phosphate and potash
- worms absolutely love it
- it has good insulating qualities.

Its disadvantage is that:

- it can harbour a range of pests.

Figure 7.24: Pea straw is a very rich mulch.

Peat moss

I am strongly opposed to the use of peat moss in any form. The bogs from which it originates are as important to the carbon cycle as are the rainforests of the world.

Pig manure

Pig manure, as indicated by its aroma, is pretty powerful material for direct use on the garden. This is because it is only partly digested within the pig and is still breaking down after it has been excreted. It is awkward to handle, it can burn a lot of plants and its odour can be offensive to the whole neighbourhood.

The only practical solution is to spread it quickly over the soil, taking care to keep it away from plants, and very promptly cover it with a dense layer of inert mulch material.

Its advantage is that:

- worms love it and it does make a good component in compost.

Disadvantages are that:

- it has a strong odour
- it attracts flies.

Pine needles

Pine needles make good mulch for acid-loving crops.

Advantages are that:

- they are high in nitrogen, with some phosphorus and a little potash

- strawberries thrive under them
- they can be plentiful if plantations are nearby.

Disadvantages are that:

- because of their high acidity they should be used only as part of a compost blend
- thick layers can pack tight and not allow air to penetrate.

Plastic

Plastic should not be laid under natural materials that you wish to become incorporated in the soil, as it presents a barrier between the mulch and the soil. Also plastic, being relatively impermeable, does not allow the soil to breathe. There are millions upon millions of aerobic microbes in the soil, especially in the top 10 cm. They need oxygen to survive and do their work of converting organic materials into the plants' diet. The plastic cuts off their oxygen supply, destroying the aerobic microbes and encouraging anaerobic microbes, which turn the soil sour.

Plastic is used in commercial growing of crops like strawberries, which do not mind sour acid soil. I suggest that one reason why such crops need repeat spraying with fungicides is the oxygen-starved soil they grow in.

Plastic weedmat is designed to overcome the breathing problem while still smothering the weeds. It should be used mainly with pebbles, to avoid the problems pointed out above. It is not suited to organic-based mulches because it breaks down not into soil nutrient, but into an untidy chemical mess.

Poultry litter

Poultry litter, based on straw and other plant material that you throw to the chickens, makes a great mulch, but it is best to place it under a protective layer of inert mulch to avoid damage by sunlight. Manure of all sorts is a beneficial addition to all compost.

Advantages are that:

- aged poultry litter contains a veritable treasure trove of plant goodies
- it is a good insulator
- it is lightweight.

Disadvantages are that:

- if it is used properly, it is hard to think of any problems except flies
- fresh litter can burn plants because of the ammonia content.

I have developed a practice of processing my lawn clippings and vegetable wastes through the chicken pen. They love it, shredding it and mixing manure into it to create an ideal base for compost and mulch underlay.

Figure 7.25: The blur of motion is a chicken claw in action – one of the best shredders around.

Rice hulls

Rice hulls in poultry litter is my favourite mulch, mainly because of convenience. I order by the truckload once a year and spread them as the need arises throughout the growing season.

Advantages are that:

Figure 7.26: Rice hulls keeping Chinese mustard moist and weed-free.

- rice hulls last and last
- they are lightweight and easy to handle
- they seldom harbour pests. My earwig population is very low, and I think that slugs and snails do not like to cross the coarse-textured irritant surface when it is dry. They will cross it when it gets wet. This, to me, is a case for drip irrigation instead of spray irrigation
- it is a natural material that breaks down in 12–18 months to form rich soil organic matter
- you can obtain a rice hull and poultry manure mix which prevents nitrogen loss
- it is weed-free (or should be, if it comes directly from the poultry shed).

Disadvantages are that:

- hulls can be partly impenetrable to water, due to a consistency that resembles sawdust, so you may wish to place your drippers underneath. Because I do not mulch thickly I place my dripper line on top of the mulch. This is because the drips soak through the mulch and the moisture spreads once it contacts the soil. To stop light penetrating, I first place two or three sheets of newspaper on top of the manure layer, then place a 5 cm layer of rice-hull mulch over that. Later, as the season warms up, I will top up the rice hulls with a further 3 cm thick layer
- hulls are flammable, but not as easy to ignite initially as straw.

Sawdust

Sawdust will do the job well but it does not allow the penetration of overhead water. So, even though in early spring it may keep the soil moist, in summer sawdust tends to prevent all but the heaviest shower or irrigation from reaching the soil.

Advantages are that:

- it insulates very well
- it smothers weeds well if applied thickly enough.

Figure 7.27: Sawdust from old timber will benefit from the addition of nitrogen.

Disadvantages are that:

- fresh sawdust will rob the soil of nitrogen, so place manure or natural nitrogen fertiliser underneath
- it is flammable.

Note that it has a fine texture which allows only a limited amount of air and water to penetrate the soil if the sawdust is laid too thickly. Keep it 10 cm away from the base of plants.

Do not use sawdust from treated pine because of the risk of poisoning.

Scoria

Scoria is crushed volcanic rock. It comes in different grades from coarse gravel to fine dust. The coarse gravel may have an appeal in some situations where gravel is appropriate. I occasionally use the dust because it can be incorporated into the soil to open it up after the dust's effective mulch life is over. Like all rock dusts it does not contain nitrogen – to aid the release of its minerals, manure, blood and bone or other nitrogenous material should be added. For coarse gravel, see p. 62.

Figure 7.28: Scoria makes a good seedbed for weeds.

Advantages are that:

- it can be used to advantage on garden traffic areas, as it will consolidate when walked upon a lot, or compressed
- it is suitable to use on flower or vegetable beds, once every 10 years as a mulch that is intended to become a soil conditioner once the weeds take over
- it is slightly alkaline
- snails dislike crossing its rough surface
- it looks attractive in many situations
- it is lightweight and porous.

Disadvantages are that:

- it is hospitable to weed germination
- it is unsuitable for repeated applications on flower and vegetable beds
- it makes the soil rough on your hands when it is incorporated into the soil.

Seaweed

Seaweed is absolutely adored by earthworms and some vegetables such as asparagus. I used it until it became illegal to gather it. Seaweed breaks down very quickly, so it is a soil improver rather than a mulch. It is extremely high in nitrogen and phosphorus and has a carbon:nitrogen ratio of 20:1, which makes it an ideal underlay for a more inert mulch.

Its advantage is that:

- it will give plants a boost as it breaks down.

Its disadvantage is that:

- it can attract flies so it is better used as a mulch underlay.

One way to get the same benefits without using raw seaweed is to water a diluted blend of liquid seaweed into mulch such as sawdust or rice hulls, or water it into the soil underneath.

Sheep manure

Aged sheep manure from under the shearing shed is a beneficial soil conditioner and plant nutrient. All manure is a beneficial addition to compost. It is best to place it under a protective layer of a more inert mulch to avoid damage by sunlight and nutrient loss.

Its advantage is that:

- it contains a good range of nitrogen, phosphorus, potassium and organic matter.

Disadvantages are that:

- fresh manure attracts flies
- it can contain harmful chemicals from the shearing shed
- it may be high in salts.

Stable manure

Aged stable manure is great. Manure of all sorts is a beneficial addition to any compost. Stable manure contains straw, horse manure and urine.

Its advantage is that:

- its texture is variable, which allows the soil some capacity to breathe.

Disadvantages are that:

- it attracts flies and you may not like the smell
- it is flammable
- the urine is high in ammonia, which burns plants. Ageing it removes this problem
- it can contain weeds.

Figure 7.29: Stable manure is a favourite of worms. It can contain cereal weeds.

Aged manure is great, but I would overlay it with straw to avoid nutrient loss to the air. It is acid so you may have to use lime.

Tanbark

Tanbark is similar to chipwood except that it can be more acid. Use it around acid-loving plants or apply lime underneath.

Its advantage is that:

- the open texture allows air to circulate. Consequently, bark mulch has to be applied 100 mm thick or you should use weedmat underneath it.

Its disadvantage is that:

- it is a fire risk.

Wood ash

Wood ash is a valuable addition to compost, applied as an underlay in acidic soils or near fruiting plants because it is rich in potash, which encourages fruit development.

Disadvantages are that:

- it can blow about, especially when you are handling it
- it is not attractive on its own.

Figure 7.30: Wood ash.

Wood shavings

Wood shavings are better than sawdust but they are a potential fire hazard and contain little nutrient value. Wood shavings from a stable or poultry shed are better.

Chapter 8

Planning your waterwise garden

One thing waterwise practice does is make you have a fresh look at your garden design, whether you have an existing garden or a new one. It pays to step back occasionally and look at the overall plan.

You can adapt the following plans and concepts to your own circumstances. The process used is a simplified version of any planning and design approach. You determine your aim, examine your resources and the site, consult all the major users, use common sense and then come up with your design. You may also include a schedule of staged development.

Your aims

Hopefully, 'waterwise garden' and 'enjoyment' will feature very strongly in a statement of your aims.

Here are a few questions that may stimulate the formation of your aims:

- Is the garden for show or privacy, or a bit of both?
- Do you favour outdoor recreation, aesthetics, or both?
- How many people live in the household?
- Does anybody in the household like gardening a little, a lot or not at all?

Make up your own questions. A statement of aim may read:

'We want a waterwise garden that the kids can play in, where we can grow a few vegetables, have privacy and not become garden slaves.'

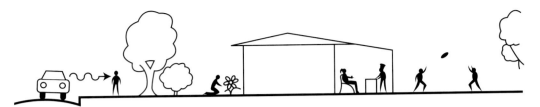

Figure 8.1: This garden illustrates the aims of privacy, gardening and recreation for the whole household.

As you proceed you may wish to modify your statement. This sample statement shows that the design should have:

- a slightly open buffer of trees, shrubs or both between your garden and the outside world
- an area of lawn that the kids can play on
- a vegetable patch
- low-maintenance planting areas
- dry and wet zones that fit in with the above aims.

Resources

Examining your resources is a reality check. Your resources consist of your finances, your available time, your attitude, available labour and equipment, and the site. Examine each and see how they influence one another.

Finances

Nobody knows your financial situation better than you. You need to work out your available funds for major items, such as purchasing and installing recycling systems or installing an underground tank.

Time

Time is a matter of priorities. You have to eat and pay your bills so you must spend time earning an income, but you also need time for recreation, socialising and relaxation.

Some time resources can be shared with the garden. For instance:

- improving the garden may add value to your property for resale
- you may share the project with members of the household. Doing something constructive together is quality time
- gardening may be your relaxation
- you may be retired and have an independent income.

Once you know your time resources you will be able to schedule any works included in the design. You will know whether this will be a long-term project or a relatively short one.

Attitude

Attitudes can move mountains – in gardening, that may mean mountains of mulch! What you value and what you really want can determine how much energy you can put into a project. If you are the sort of person who sees waterwise gardening as a worthwhile challenge then you are more likely to be motivated to make the available resources fit into achieving that challenge.

A little self-honesty at this stage can save a lot of frustration later. If you are inclined to be a dreamer rather than a doer, tailor your waterwise endeavours accordingly.

Labour and equipment

Labour consists of the number of enthusiastic helpers, their skills and your ability to co-ordinate. Coerced workers are an impediment.

Site assessment

Site plan

If the garden is a new one, the site assessment is fairly simple. You will need a 'site plan'. Use photocopies of the builder's site plan. I suggest that you make five or more copies at A4 size, for rough notes and to allow for changes or errors, and a couple at A3 size, for your site assessment summary and the final design plan.

If you have an existing garden then you will need to include it on your site plan. You may have the original plan, or you can obtain a copy of your site plan from your water authority or your council.

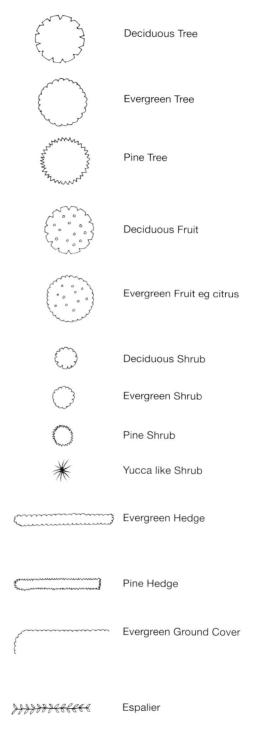

Figure 8.2a: Sample symbols of plants, that can be photocopied and stuck on plans.

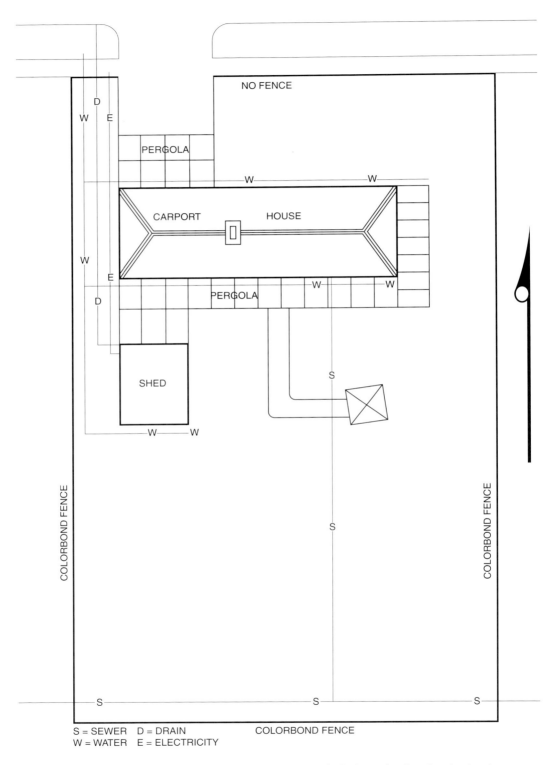

Figure 8.2b: A site plan based upon the best plan you can find of your land and main structures.

Your plan should include:

- all buildings and structures
- services above and below ground including water services, sewerage lines and sprinkler systems
- paved areas
- existing garden features including lawns, beds, rockeries, vegetable patch, clothesline, poultry run etc.
- anything that may affect your design or be affected by it.

How to proceed from here

You do not have to draw all the assessment plans that follow. They are mainly for demonstration purposes, but you should understand the principles involved so that you can draw the final assessment plan. Of course you may wish to draw up each stage: the choice is yours.

Your garden climate

The plan below is designed to give you an idea of a typical set of garden climates, so that you can make similar judgements on your property. The climate of your garden really consists of several different climates, or micro-climates, that have seasonal patterns. A sheltered part of the garden has a more moderate summer and winter than an open exposed part of the garden. Figure 8.3 shows a typical set of micro-climates within a suburban garden without trees and shrubs. Summer temperatures in shaded areas remain below 40°C. In exposed areas they soar to above 50°C.

It is important to know the micro-climates so that you can grow plants that will survive, and to arrange plants that modify extremes for more fragile plants.

Micro-climate zones (circled)

1. Exposed to the north and sheltered from the south; sun most of the day. North winds in summer. Summer temperature up to 50°C.
2. Morning shade; afternoon sun; winds from the south, north and west; sheltered from the east.
3. Morning sun; afternoon shade; winds from the south, east and north.
4. Midday sun; north and south winds funnel through and are thus accelerated.
5. Midday sun; north and south winds funnel through.
6. Shade for most of the day and year; temperatures up to 40°C in summer.
7. Morning sun; some north and south wind.
8. Sun most of the day; some wind from all directions; summer temperatures up to 50°C.

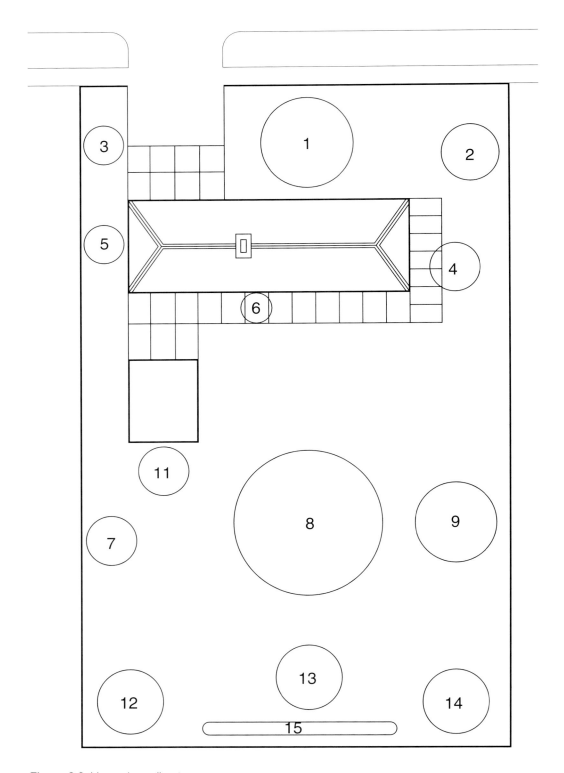

Figure 8.3: Your micro-climates.

9   Sun from midday; some winds from all directions except east.

10 Midday sun; north and south draughts.

11 Shade most of the year; sheltered from the north.

12 Morning sun; sheltered from the south and west.

13 Sun most of the day; moderate winds from the south, east and west; stronger winds from the north.

14 Shade in the morning; sun from midday; sheltered from the south and east.

15 Sun most of the day; mostly sheltered except for winds from the north.

You can see immediately that the climate pattern in an area without trees and shrubs will alter dramatically once trees and shrubs are added. The whole picture becomes more complex, but we can safely say that the garden climate will be moderated considerably.

The pattern of micro-climates might remain the same but their extremes would be greatly moderated. Dense planting would change the micro-climates completely, where a light planting would only moderate them. You must make your own garden climate inventory to enable you to plan your drought containment strategies.

The main house and sheds cast a daily cycle of shade. Remember that as the sun gets higher in the sky toward midsummer the shade band on the south side of structures can narrow to nothing.

It is important to know how micro-climate is affected by shade, daily and seasonal temperatures and wind patterns, and how extremes can be modified by structures. These effects are demonstrated in the next few pages.

Shade

Experience has shown that plants grown in partial shade throughout most of the day are healthier and more likely to survive drought.

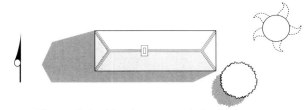

Figure 8.4a: Morning sun and shade.

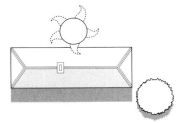

Figure 8.4b: Midday.

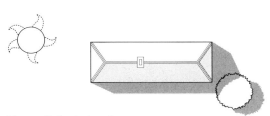

Figure 8.4c: Late afternoon.

Temperatures

Temperatures in summer can be severe and prolonged.

Figure 8.5a: Morning is generally the coolest part of the daylight hours. In frost-prone areas it is the most damaging time – the sun reaches exposed areas such as lawns, warming them too suddenly and causing stress.

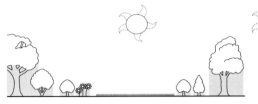

Figure 8.5b: Summer temperatures in unshaded areas can reach 40–50°C at midday, lasting into the afternoon. Lawns, in particular, suffer. Shaded areas can be up 10°C cooler.

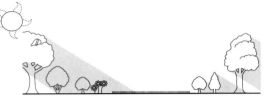

Figure 8.5c: Summer temperatures in exposed areas can remain hot into late afternoon, but increasing shade has a moderating influence.

Wind

Strong wind is the enemy of waterwise gardeners because it increases evapo-transpiration. Gentle breezes are essential for healthy plant growth, because they reduce the risk of problems such as fungal infections.

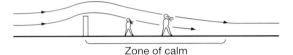

Figure 8.6a: The effect of a solid barrier such as a brick fence upon wind patterns (not to scale).

Figure 8.6c: The effect of a combination of vegetation and constructed barriers allowing gentle air circulation but reducing most of the force of strong winds (not to scale).

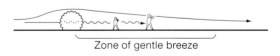

Figure 8.6b: The effect of a semi-permeable barrier such as vegetation (not to scale).

Gentle air movement can be achieved with sensitive planting and fence construction. It is important to understand how wind reacts to different barriers.

When wind at ground level meets a solid barrier such as a fence it deflects upwards and returns to ground level at a distance past the fence approximately equal to five times the height of the fence. There is little air movement close to the downwind side of the fence, but it increases further away from the fence. Dense vegetation can achieve a similar effect.

With the same wind, a semi-permeable barrier such as open vegetation or a picket fence will allow gentler breezes through but stops the main force of the wind at ground level. It will take up to 13 times the height of the vegetation or picket fence for the wind to return to ground level.

The aim is to achieve a combination of fences, buildings and vegetation that moderate the stronger winds so that you get the breezes you want, especially on hot days.

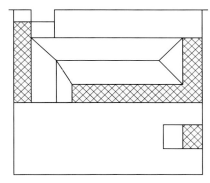

Figure 8.7: Good use can be made of pergolas.

Pergolas

Pergolas cast an almost ideal shadow. They moderate sunlight and breezes. Each pergola will be different according to its location and construction. Pergolas can be excellent zones for sowing pot plants and propagating, especially if near the house.

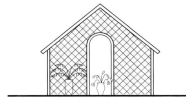

Figure 8.8: Economically constructed shade areas.

Gazebos

Gazebos are similar to pergolas except they do not allow rain or overhead sun to penetrate. Any plants within a gazebo will need to be watered.

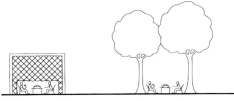

Figure 8.9: Trees – even advanced trees – are cheaper than any structure.

Ferneries

Ferneries can be temporary homes to many pot plants and good places to start off seedlings. Remember that when plants leave the fernery they will require hardening off.

Topography: drainage

Ideally your plan should show contours, but this is not always possible. The main thing is that you should know the direction of drainage on your land.

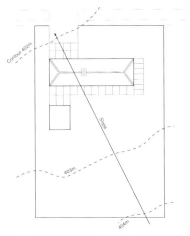

Figure 8.10: Drainage directions shown by arrows. The drainage line runs at right angles to the contour lines (shown dotted).

Existing plants

You may consider it good enough just to locate the existing beds and a few specimen trees and shrubs on this plan. Since it is very unlikely that you will be shifting anything mature, aim to design the garden around them.

Impediments and assets

Impediments and benefits to your quality of life, whether originating on- or off-site, should be taken into account when planning your garden. These include your need for privacy, noise abatement and wind reduction, and minimising fire risk. Use symbols to show these on your assessment plan.

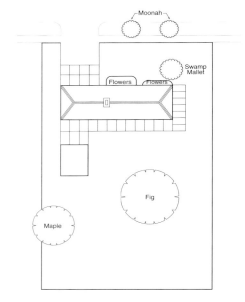

Figure 8.11: Existing vegetation.

Privacy
Privacy can be achieved by planting vegetation strategically. The sorts of things you may like to screen out are pedestrian and vehicular traffic, neighbours and any tall buildings overlooking your land. These should be noted on your site assessment by a symbol that you can easily recognise.

Noise
Noise is another factor that can be reduced by vegetation, but don't be too optimistic about its effectiveness. A solid brick or stone wall or an earthen bank is far more effective for noise abatement than foliage.

Wind
Wind can be reduced considerably by vegetation. On your site plan, note the directions of the prevailing winds throughout the year. You can get these details accurately from the Bureau of Meteorology. Armed with this information, you can decide which sides of your land need to be planted out. A word of warning: a gentle breeze may be desirable in hot seasons. So be careful not to plant out the side that lets in summer relief. See also p. 88.

Views
Views can be good or bad. Your garden design task is to screen out the undesirable ones and open up any desirable outlooks. This may include an old shed that you wish to hide. Mark both good and bad views on your assessment plan with symbols that you either copy or invent, and are easy to remember.

Fire

It is not my intention to alarm you, but certain drought-tolerant trees and mulches are very flammable. The main offenders are rough or loose-hanging bark eucalypts or the many wattles with low-growing foliage. Cypress hedges are most flammable, as many victims of suburban firebugs can attest.

Fortunately, good planning can reduce the problem to a minimum through species selection, strategic placement of fire-retardant species and lawns.

For the purpose of the site assessment plan, you need to show the direction of maximum summer risk. This can be the direction of strong winds, the side where there is volatile neighbouring vegetation, neighbours' open wood-fired barbecues, or trees that overhang your house.

For your convenience I have listed many fire-resistant plants suitable for fire buffer zones in Appendix 4.

By now your site assessment plan may be looking very complex to strangers, but hopefully not to you.

Figure 8.12a: High-risk effect of volatile vegetation compounded by the weatherboard house with a gap at the base that would allow the entry of embers.

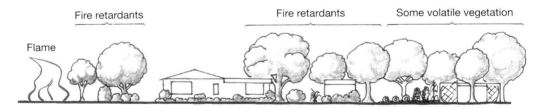

Figure 8.12b: A fire-safe design.

Site assessment plan

The guidelines we have just shown to help you prepare the site assessment plan are the same guidelines a professional landscaper should follow for any client. It is your tool to help you complete your design plan. See Figure 8.13.

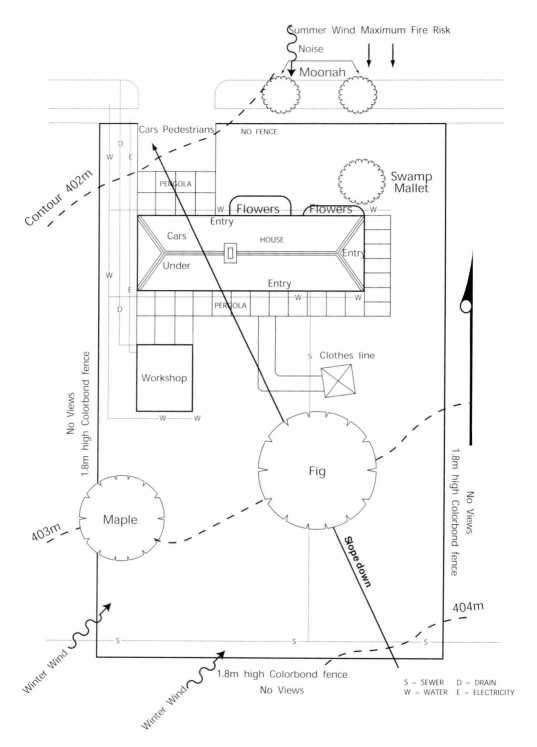

Figure 8.13: The final site assessment plan, summarising all the factors influencing the design for the site.

Design

Your design plan will reflect your aims and take into account the site assessment plan you have just prepared. It should last for the duration of your project, which may be a few months or years. But it need not be permanent, and can change with circumstances.

Activity areas and linkages

On a site plan showing just the house and outbuildings, circle with pencil the various activities you wish to carry out in your garden. Then show the linkages between the various activity areas. Linkages are simply the way you would walk between the activity areas. They will indicate where you should have openings between plantings to allow for paths and drives.

Dry zones and wet zones

Using your site assessment plan and your activity plan as guides, you can now designate dry zones and wet zones.

Dry zones
Dry zones are where you will apply little water. They include native, Mediterranean and desert plantings, along with paved and gravelled areas.

Wet zones
Wet zones are where you will plant water-loving plants and set out your watering systems, including sprinklers to serve lawns, shade areas, ferns, vegetable plots and a few of your favourites that you like to spoil.

The object is to make the wet zones as small as possible without detracting from your enjoyment of the garden.

Wet zones are generally closer to the house with most of the dryer areas further out, mainly because this approach is more economical in placing water systems. Of course some wet zones such as vegetables may be at a distance from the house and the main dry-zone paving is generally closer to the house.

For and against lawns
Lawns are a delightful feature that we have inherited from Europe where it rains a lot. All the old lawns around stately homes were established with little in the way of installed irrigation systems. In our sometimes-harsh climate, the use of lawns is subject to debate.

The case against lawns is based on the fact that they are generally the largest suburban users of water.

There is, however, much in favour of lawns.

- Lawns look good and set off the house and garden features to their best.
- We play games and relax on lawns.

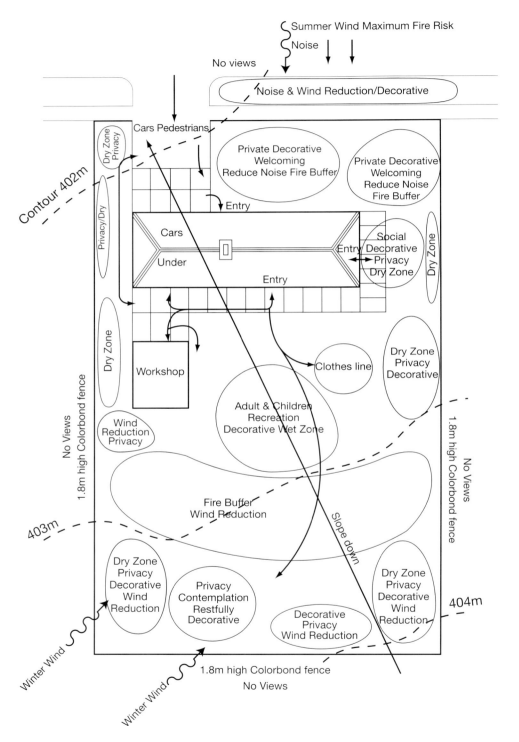

Figure 8.14: Activity areas circled. Linkages indicated by arrows tend to be pathways or driveways.

- Lawns supply much of our mulch and compost material.
- Lawns are a good buffer against fire.
- If you keep poultry, lawn clippings help to make the egg yolks richer.

The trick is to use lawns strategically so that they serve the good purposes, but not to go to excess. There are new lawn blends on the market that do not need as much water as the older varieties. Your local water authority should be able to advise on suitable lawn species.

Fire buffer zones

Fire buffer zones are required only where there is a lot of bushland near your property. In other areas it is enough to keep flammable plants at a sensible distance from the house. These zones consist of materials that do not burn readily and act as barriers to flame:

- Paved or gravelled area.
- Lawns.
- Fire-retardant plants, which do not burn readily or do not burn at all. See Appendix 4.

Examples of wet/dry zone fire-retardant designs

Figure 8.15: A golf course is a perfect example of the dry zone/wet zone principle.

The following are examples of differing approaches to designing dry and wet zones with fire-retardant plantings in various situations. They can be adapted to individual tastes and budgets. They can start off simple and be added to from time to time.

Figure 8.16a: This site is relatively flat and the dimensions can vary in width and length.

Figure 8.16b: This site is flat, or slopes slightly up or down away from the viewer.

Figure 8.16c: This site is flat or could be stepped slightly up or down halfway along the pathway. The location could be tropical, subtropical or temperate.

Figure 8.16d: This site can be flat, or rising or falling away from the viewer.

The areas shown are adaptable. Think of them as viewed from the house looking toward the front fence, or back fence. Some can be adapted as side gardens.

The design in Figure 8.16a has an Arabian influence. It features a border of espaliered fruit trees and low hedges. The centrepiece is a lawn in the shape of a Middle Eastern pool. The focus at the end can be a statue, an urn or other feature. The lawn is surrounded by gravel or paving.

The design in Figure 8.16b reflects the English formal influence. The formal trimmed hedges may be dry-zone species, water-lovers or semi-hardy, sharing some of the lawn watering. The central dividing path could be paved or gravel. This simple design lends itself to adaptations involving low border hedging around herb or flowerbeds. The lawn may be the only wet zone.

The hedging shown in Figure 8.16c can vary in height and foliage density according to the need for protection from strong winds. The hedges may be trimmed and formal, or random shrubbery. The lawns are the only wet zones. Path surfaces are optional.

Planning your waterwise garden | 99

Figure 8.16e: This site can be flat or rising or falling away from the viewer.

Figure 8.16f: This layout suits a terrain similar to the one in Figure 8.16d but on a more extensive site. The levels vary and the wet zone/dry zone plantings are optional, with an open area near the house.

The design in Figure 8.16d is less formal, with open activity areas surrounded by shrubbery. Lawn or paving can be used. The stairway linking the two areas is necessary only if there is a change of level. If the site is flat or slightly undulating, the link could become a pathway or lawn linking the two open areas, possibly including an archway to give the feeling of moving from one outdoor room to another. The open areas can be lawn (wet zones) or paving or gravel (dry zones). Plantings can be grouped in fire-resistant bands around the open areas. The wet zone or fire-retardant species should occupy the foreground closest to the house.

The design in Figure 8.16e is simpler than that in Figure 8.16d, and has a more rectangular feel. The wet zone can be lawn in the foreground, complemented by an arid zone planting. The background can be lawn, paving, gravel or a vegetable patch in the open area surrounded by dry-zone planting.

In all these designs privacy can be enhanced and the impact of wind reduced by selecting taller species for the boundary plantings. Open areas can be recreation areas using appropriate surfacing.

Model kitchen garden

The following model can be used as a wet zone in most designs.

Ideally the beds should run north–south to get the full benefit of the sun. The drainage sump is equipped with a small pump to recycle nutrient-rich drainage water, which would otherwise enter and pollute the groundwater. The subsurface drains in Figure 8.17 are shown as broken lines beneath the paths leading to the recycling sump from where the drainage water is redistributed to the fruit trees. More detail is shown in Figure 8.18a and Figure 8.18b.

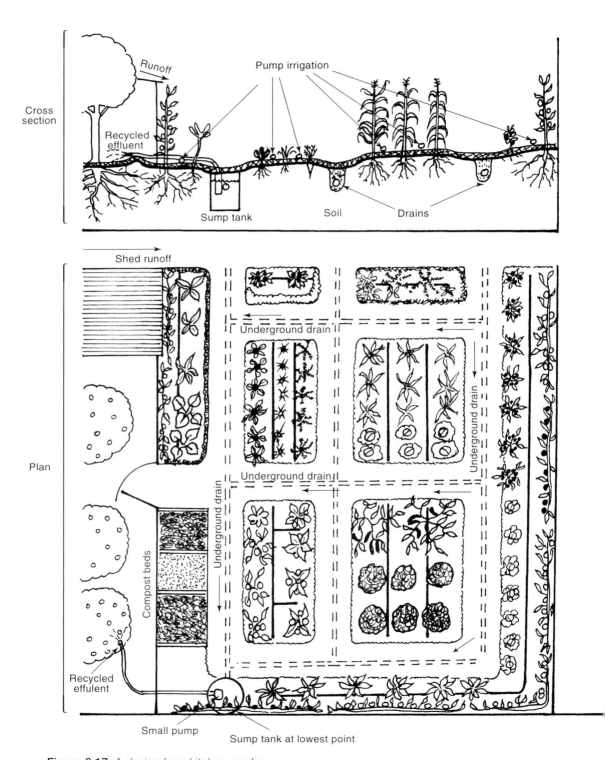

Figure 8.17: A design for a kitchen garden.

Garden water recycling

One area of modern gardening that has hardly been explored is garden water recycling. Many attractive gardens are overwatered, particularly lawns and vegetable areas. In many cases, subsurface drainage systems are a part of good garden design and management. The question is, what should be done with the drainage water? Take it to the lowest part of the property then send it somewhere else?

A few further questions need to be answered.

- What is the quality of the lost water?
- How much is lost?
- Where does it go?
- What is its impact?

We have already looked at environmental impact (see p. 9). My guess is that 50% of water on the wet zones of gardens ends up in groundwater. This estimate could be exceeded where timers and waterwise practices are neglected.

In a healthy garden, drainage water is sure to be nutrient rich – this resource can be lost along with the water. The answer to the question of how much is lost depends upon variables such as watering practices, appliances and duration of watering. Soil type and depth also play a part, as do seasonal conditions.

One thing is certain: a lot of garden water goes to waste and it is likely to be a major source of water and nutrient loss.

Home trial

To help answer these questions, I am undertaking a home trial. The methodology is not strictly scientific, but it will provide a guide. Figure 8.17 gives an idea of how the system works in a kitchen garden. The illustrations below give details of the system.

The cross-section shows a perforated 80 mm diameter polythene pipe under the path between vegetable beds.

The pipe in Figure 8.18b forms part of the drainage system leading to the underground sump at the lowest point in the garden, as shown below. The sump being tested is a long-life plastic 100 L capacity drum. It must be made childproof.

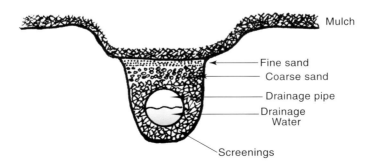

Figure 8.18a: Cross-section of the drain set below the path of the kitchen garden.

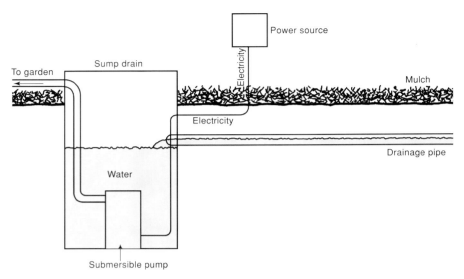

Figure 8.18b: Detail of the drainage sump. The drain empties into the sump drum, which has a submersible pump to recycle water back around the garden.

The power source could be mains supply or, as in the trial, a jet venturi assembly. The principle of the jet venturi is a little like that of the detergent cylinders that you can attach to the hose when washing the car, as it operates on the power of water delivered through the hose. The result is a cocktail of mains water and whatever is in the cylinder. Our cylinder is an old 44 gallon drum containing the nutrient-rich drainage water. The cocktail then consists of a blend of mains water and reused water. Determining the desired ratio of mains to reuse water is part of the trial.

The jet is capable of creating a lift of a few metres below that of the mains supply that drives it, which should be adequate for most sites. The outlet hose is 25 mm polythene pipe leading to a flexible standard garden hose, which can be moved to water different locations. In the trial the hose will be used to water fruit trees, compost and worm beds. It could even be used on vegetables if the right sort of spray or dripper can be found.

The trial system is designed to cope with 400 L of reuse water a day. A storage tank may be included for temporary excess flows.

Although the system can operate all year, it may be turned off during wet seasons.

The trial is not yet sufficiently advanced to draw final conclusions. If you decide to undertake your own trial, please let me know your results. The results of my trial will be included in my forthcoming book on organic gardening, *Eat Your Garden*.

The master plan

The master plan is a detailed design based on the processes that you have followed to this point. You will use it as your guide in developing your waterwise garden.

Figure 8.19a: Example 1.

A typical garden design includes:

- In fire-prone areas, avoid the use of shade-cloth and wooden trellis in extensively constructed and planted shade areas near the house.
- Limited lawn, acting as part of the green fire-retardant band around the house.
- Paved areas.
- Dripper watering systems.
- Drought-hardy plantings.
- Wind barrier plantings and constructions.

Apart from the extensive use of shade trees, the garden looks like a conventional garden. This is because of the versatility of fire-retardant plants.

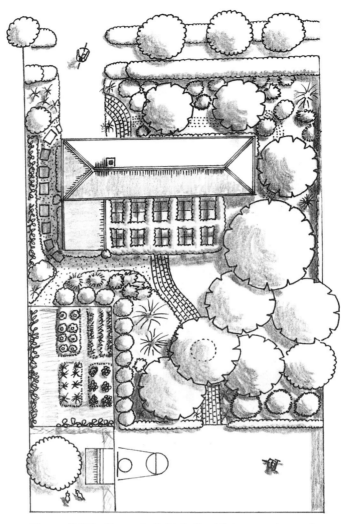

Figure 8.19b: Example 2. Note the kitchen garden wet zone, the poultry run and the recreational paved area at the very rear. This garden caters for a different set of household needs from those in the first example.

You may have to stage your development if:
- the project is a large one involving a lot of changes
- costs are high
- you have to take seasonal conditions into account
- there are major constructions
- your time is limited.

You can create your own timeline to take into account of these impediments.

Chapter 9

Plants

Your prize plants

Some plants will expire more quickly than others when left without water, and the tender ones are quite often your prize plants. You may be surprised to learn that all plants can become drought-hardened to a certain extent, some more than others.

If water in the soil is partially reduced the plant uptake is less and growth is restricted as the plant toughens itself on a reduced uptake of water and nutrients. Leaves, blooms and fruit will not be as lush and abundant – although in some plants blooms are more prolific as a stress signal – but the plant survives. If you want the fruit to survive you will have to increase the water supply once blossom appears. If the dry is extreme and water is scarce, the tree will survive better if you remove fruit.

You can find the survival level by observing each plant as water supplies are reduced.

Deep mulch and shade are also your allies with these plants. If there are no shade plants nearby you may have to build a shade-cloth frame.

Monitor plants under stress for pest or disease attack. If possible, shift tender plants to sheltered areas, but do not dig up mature plants for transplanting or potting. They will suffer additional transplant stress.

The exceptions to this rule are vegetables – these do not like moisture stress at any time.

Survival plants

Over time, plants have developed various strategies to reduce the effects of drought, wind and sun. These include:

- desert dormancy (remaining dormant until rain falls)
- shiny upper surfaces that reflect sunlight

- furry silver leaves that reflect heat, while the hair reduces evapotranspiration due to wind
- leaves that hang vertically and are in partial shade most of the time – some plants have leaves that actually turn with the sun, exposing the edge only
- leaves high in oil content, that require less water
- broom foliage instead of leaves, reducing surface area subject to evapotranspiration
- spiky foliage such as in cacti for the same purpose
- deep tap and other roots that seek moisture at lower levels
- succulents can store great amounts of water in fleshy leaves or stalks. Various desert-dwelling trees store water in bottle-shaped trunks
- growing in the shade of taller companion plants.

A detailed description of plant survival strategies is given in Appendix 1.

Australian plants

As you would expect, many of the 15 000 Australian plants have evolved in a drought regime. That does not mean all Australian plants are dryland plants – Australia also has tropical, monsoonal and temperate rainforests. Plants that have evolved in the wet coastal fringe and on the slopes of mountain ranges are less equipped for drought than those that are found in the mallees and deserts.

Many rainforests and wet sclerophyll forest plants are too big for our gardens anyway. Few people have enough room for mountain ash and blue gum, both of which require annual rainfall of over 1000 mm to grow vigorously. Blue gums with their vigorous early growth have, in the past, been popular with novice gardeners or in farm shelterbelts. But the early growth gives way to disease, insect attack and death as a result of water stress. The plains of western Victoria are dotted with plantations of dead blue gums, which initially grew well but failed over time due to moisture stress.

Waterwise gardeners should look at the wide range of dryland eucalypts. Mallees are ideal in size and form for the home garden. Flowering gums such as yellow gums and ironbarks can grace larger gardens, away from buildings. The Western Australian flowering gum, *Eucalyptus ficifolia*, is ideal for most gardens. *Eucalyptus caesia* is spectacular. Not only are the flowers dramatic, but the bark adds interest year-round.

Acacias
Acacias (wattles) are good drought value, although some are short-lived. The range of choice is enormous.

Casuarinas (and Allocasuarinas)
Casuarinas form a very useful class of drought-tolerant trees with their fine leaves and capacity to self-mulch. A copse or even a garden of casuarinas, with its carpet of soft needles, does not need lawn.

Callitris
Mallee pines love good drainage and can survive with around 200 mm of annual rain.

Banksias and grevilleas
Banksias and grevilleas are some of our hardiest plants. They prefer near-dry, well-drained conditions and should not be watered or fertilised. Fertiliser and too much water can kill them.

Callistemons and melaleucas
Callistemons and melaleucas can tolerate either wet or dry conditions for fairly long periods. They are the native equivalent of pines or cypress.

Correas
Correas with their bell-like flowers like well-drained conditions although they can tolerate relatively heavy rainfall.

Hakeas
Hakeas can be found in hot dry situations. Some, such as *Hakea francisiana*, thrive on heat.

Dillwynias, bossiaeas and chorizemas
Dillwynias, bossiaeas and chorizemas are dry-tolerant plants with decorative pea-like flowers in spring.

Grasses
Many of our grasses have adapted to dry conditions, especially in the dry north. You may need to develop an appreciation of the dry foliage of such grasses as kangaroo or speargrass. In general, native grasses form coarse lawns or tufts not suited to heavy traffic. They are obtainable from specialist nurseries. Consult your local water authority or botanical garden for advice on decorative grasses for your site.

Myoporum
Myoporum parvifolium is a great groundcover in non-traffic areas. It is also fire-resistant.

Hardenbergias and kennedias
Hardenbergias and kennedias are groundcover or climbing vines with colourful flowers.

Plants of the world

Plants from Mediterranean or desert climates provide culinary, decorative and medicinal dimensions to the garden.

Rosemary and lavender are great drought-tolerant mainstays. Lavender oil quality even improves if the plant is not watered artificially. Lavender and rosemary both get by well with just winter and spring rains.

Sages and thymes are adapted to dry conditions but like a little water occasionally. These herbs like drainage and prefer slightly alkaline soil conditions.

Olives, conifers and palms complement the herbs and give an exotic effect.

Agapanthus from Africa is a dramatic blue or white summer addition, but harbours snails.

Rosemary and lavender store oils in their foliage, purslane is a succulent that stores water.

Most of the exotic plants listed in Appendix 3 may not be as drought-tolerant as the Australian plants in Appendix 2, but they are the best the world has to offer us. Remember that no list is ever complete and you should supplement it with your experience. Heights of plants shown are a guide only because plants vary according to soil types and climate. Many exotics have an advantage over most Australian plants in several ways.

- They are fire-retardant. Deciduous and succulent plants hold water in their foliage and, being less likely to burn, can be part of any fire-buffer plan.
- They create shade. Many exotics have dense foliage and cast heavy shade.
- Many bear fruit.

Chapter 10

Help with plant selection

In most major centres around Australia, there are good native plant nurseries and demonstration gardens in botanical gardens and elsewhere. The plant lists from many of these sources are given (see Appendices 2–4) so that you can make your choices.

Figure 10.1: Basaltica demonstration garden displaying waterwise plantings suitable to the volcanic plains west and south-west of Melbourne. The garden was inspired by Ross Mellor OBE, designed by Robert Harvey and funded by CRT and City West Water. It is located at CRT headquarters at Altona. The area is promoted by City West Water.

Your local water authority

Your local water authority could be one of the best sources of advice on the plants to use in your garden. Many water authorities have researched the best plants for their district and display them in local waterwise gardens. They may also have independent advice on such things as the recycling technology and water tanks best suited to you.

Your local indigenous nursery

There are indigenous nurseries in most districts, and it is a good idea to shop around for the best-quality stock. Although price is a consideration, you should concentrate primarily on quality. There is nothing worse than looking after a plant for a year only to find it dies or gets blown over.

You can identify a good nursery by the quality of its stock. You can get a very good idea by closely examining the tube or potted stock. If plants are too advanced for their container their roots start to spiral. This spiral growth continues when the seedling is planted out and begins to mature. In many cases a strong wind will blow over or snap off the poorly supported plant.

Figure 10.2: Bendigo's indigenous plant nursery, Goldfields Revegetation, is the brainchild of proprietor Marilyn Sprague whose encyclopaedic knowledge and enthusiasm inspires Central Victorian waterwise gardeners.

Your local botanic garden

The botanic gardens in major cities have waterwise demonstration gardens. For example, the Brisbane City Council has a display of arid-zone plants at its Botanic Gardens, Mt Coot-tha Road, Toowong.

Figure 10.3: Waterwise plantings at Royal Melbourne Botanical Gardens.

Public parks and specialist gardens

Check with your local council for the location of gardens with indigenous plants. Your state 'Open Garden' scheme involves different gardens on display throughout the year. For example, Townsville City Council has a magnificent Palmetum where many plants are not irrigated.

Societies for growing Australian plants

There are several state-based societies, using names such as Australian Plant Society (Vic., NSW) and similar names.

ACT	Society for Growing Australian Plants Canberra PO Box 217, Civic Square ACT 2608
NSW	Australian Plants Society PO Box 744, Blacktown NSW 2148
Qld	Society for Growing Australian Plants PO Box 586, Fortitude Valley Qld 4006
SA	Australian Plants Society PO Box 304, Unley SA 5061
Tas.	Australian Plants Society PO Box 75, Exeter Tas. 7275
Vic.	Australian Plants Society PO Box 357, Hawthorn Vic. 3122
WA	Wildflower Society PO Box 64, Nedlands WA 6909

On-line

The Society for Growing Australian Plants has a number of web addresses.

NSW	www.austplants-nsw.org.au
Vic.	www.vicnet.net.au/~sgapvic
SA	www.iweb.net.au/~sgap
Qld	www.sgapqld.org.au
Tas.	www.wilmot.tco.asn.au/apst
WA	www.ozemail.com.au/~wildflowers

Various water authorities are listed in your phone book, many advertising their website address. Website details are also given on your water account.

Chapter 11

What to do during a drought

Drought can be a depressing time for gardeners. Watching plants you have nurtured suffering stress and dying is not the most pleasant experience. Your best hope is to develop a positive approach. Save the plants that you can and learn from the experience so that you can prepare for future droughts. If you have followed most of the advice in this book you will have already adopted a positive approach and you should come through a drought with most of your plants intact. Your garden will be ready to leap ahead as soon as better times arrive.

Every Australian gardener is likely to experience drought several times during their gardening life. Establishing a garden is hard and costly work. So is re-establishing a garden.

Preparing for the next drought calls for good drought-proof design and plant selection. If you have not built or bought your house yet, use the principles shown here to help with your choice of land, garden and house design, and plant selection.

Essentials

Water supply
Maximise the supply of water to your property. Whether it is a piped supply or otherwise, ensure you take full advantage of the pipe gauges permitted and the quantities your supply authorities allow. On our 8 ha we get the maximum water issue available from the channel supply.

Maximise your on-site water storage, even if it is water tanks filled from the roof of the house and shed. Make great use of water economisers such as drippers and soaker hoses, and minimise mist sprayers.

Lawns
Minimise the area of lawns. Even in Europe, lawns are usually limited. Try one of our native grasses or the new low-water-use grasses on the market. They are much better equipped to survive.

Shade
Shade is a major factor in reducing evaporation at ground level. A canopy of native, low-transpiring trees, or selected exotics, can transform your place from a desert of despair during a drought to a cool haven where some green remains.

I shall never forget travelling during a desolate drought in Queensland, through mile after mile of dust and despair, to come upon one farm property practising agroforestry. Not only was there grass under the canopy of trees, but some of it was actually green. This lesson can be applied to gardens Australia-wide.

Plant selection
Observe what grows naturally in your area, and use it. Consult your local native nursery specialist. Visit your local herb nursery. Consult the lists and sources in this book.

Develop a routine

Develop a watering routine based on the plants' requirements and the water available. Some plants will require daily attention, some weekly, some monthly and some hopefully, not at all. Individual attention to plants will help the observation and learning process.

Start the waterwise design process

You should start planning during the current drought for the next drought, based upon your experiences and observations.

Draw a plan of your garden climates. Develop shade and mulch strategies to improve the climate.

Select construction projects

Since you will have time free of mowing, why not use it to build a pergola, pave an area or fence a windbreak? Construction activity is food for the soul.

Determine not to forget

There is a drought-resistant plant for most situations. However, not every plant you select need be drought-resistant. You should cover at least your essential food requirements. Fruit trees, berry bushes and vegetables should have a place in everyone's garden. Try to locate them fairly close together so that a watering system can be devised to cover them efficiently and that appropriate mulching with manure, hay or the like can be applied to your productive area.

Use drought to your advantage
Create a drought defence perimeter around your vegetable garden or other tender plant zones to exclude pests.

Although snails and slugs may perish during drought, their survival method for the next generation is to leave eggs in crevices and cracks in soil, bricks and timber. So remove timber, bricks and so on from the perimeter. Use gravel for paths. Rough materials such as scoria are ideal. Snails are loath to cross dry rough expanses.

Include your poultry run or dry native plant zones in the perimeter. The chickens will consume anything that tries to cross.

Look after the animals

Birds, mammals, reptiles and even insects all suffer because of drought. The very least you can do is to put out a bowl of water and keep it filled for the length of the drought, because the animals will come to rely upon it. Consult your local field naturalist club about what animals could be in your neighbourhood and what to feed them.

Since the drought commenced around Bendigo in 1994 we have had wildlife coming in to feed on our small areas of irrigated lawn. It is surprising how thirst and the need for food make animals tolerant of humans. Ducks, ibis, cranes, small birds and kangaroos all visit regularly from the nearby forest. Even a previously unsighted wallaby, which is usually territorial, has extended its range to include our violets, grapevines and lawn.

Figure 11.1: Help your local birds.

You can help many of the birds in your area by growing nectar-producing plants such as callistemons and grevilleas.

Spiky plants such as *Grevillea rosmarinfolia* offer birds both protection from cats, and a source of food.

Things to try

Cut the area of your vegetable plantings by half, but don't waste the remaining half. Grow pumpkins there, using the bucket-watering method at the base of the vine. Plants in the ground can get by with less water if mulched well and watered wisely.

Using pots to advantage

Pots, especially absorbent terracotta or concrete pots, are great users of water because they drain easily. But you can turn this situation around, especially with lighter plastic pots.

Place pots in shaded areas near the water source. Shade will reduce water use and you won't have to cart the water so far.

Put a dish under the pot. When it fills with drainage water, recycle the water back into the pot. You will also be returning soluble nutrients. Do not let your pots sit in water, as this will cause root disease.

Another method is to observe the base saucer during watering and to cease watering once water starts to filter into the dish. Finally, give the pot plants a heavy prune to reduce water use. Remove up to half the foliage.

You can transfer your vegetable plot to pots next to the water source. I advocate a reduced vegetable plot in most situations because lifting pots can become a burden.

Try oak leaf lettuce or cos lettuce from which you can pick leaves. Such plants are more productive for a longer time in a small space.

Deep mulch and shade are also your allies with tender plants. If there are no shade plants nearby you may have to build a shade-cloth frame.

Watch plants under stress for pest or disease attack. If possible, shift tender plants to sheltered areas, but do not dig up mature plants for transplanting or potting. They will suffer additional transplant stress.

Golden rules for the waterwise gardener

- Think Australian.
- Think Mediterranean.
- Think desert.
- Think dry.
- Forget European.
- Forget about trying to emulate a rainforest, unless if you live in or near one.
- Forget big-leaved plants.
- Forget about planting annuals, except those that self-seed.
- Mulch, mulch, mulch.
- Shade is vitally important.
- Use companion shade plants to protect smaller plants.
- Think fire prevention and buffers.
- Combine any oasis with your greenbelt of fuel reduction zones.
- Forget about having extensive lawns.
- Use native grasses for lawn areas.
- Set a reduced water consumption target of less than half your previous one.
- Don't use sprinklers except in the fire protection zone.

- Use drippers, but only where absolutely necessary.
- Design house drainage for water storage or to tap into with a hose.
- Ensure that the garden is well drained – most drought-resistant plants do not like extended periods with wet feet.
- When buying property, consider access and sites for water tanks.
- When property-hunting look for an area free of pollution and a roof suitable for water collection (i.e. no clay tiles) to allow the option of installing a water tank.
- When building, ensure your mains connection is the largest gauge possible under local regulations.

Enjoy and treasure your garden. You are one of the lucky one in a hundred people on this planet to own a garden.

Chapter 12

The future

In this chapter we discuss the practice and philosophy of water use at a community-wide level.

Black water perspective

Although this book is mainly about waterwise efforts at home we should also review our attitudes and practices toward the more complex question of sewerage and drainage at a broader level.

In most households, sewerage and drainage is not given much consideration, but there are three aspects the waterwise householder needs to consider. The first is the quantity of water being used to transport our wastes; the second is the value of the waste; and the third is the impact of our waste disposal practices on our wider environment and economy.

The design, installation, maintenance and replacement of our sewerage and drainage systems are a sizeable part of our government's capital outlay and it is useful to see how these systems have evolved. The perspective of history gives clues and motivations about how we can deal with the current situation in our communities and homes.

Modern cities treat sewage as waste. In the past this was not always so. The Chinese, for example, have been recycling sewage for four thousand years to sustain their agriculture. They, the Japanese and Koreans, apply nearly 100 million tonnes of sewage annually to agricultural lands. The early Romans and later Europeans used sewage as soil improvers. Rome was the first sizeable city to develop an underground sewerage system.

An 1884 US government report entitled *City to Farm: Urban Wastes and the American Farmer* stated: 'all human and animal manure which the world loses, restored to the land in stead of being thrown in water, would suffice to nourish the world'. But when modern communities developed sewerage and drainage systems to improve sanitation and prevent diseases, sewage was treated as a waste product to be disposed of. Our homes and streets became clean and pleasant places with fewer diseases and smells. The unpleasantness was transported to sites remote from our communities, using water as a means of transport. This technology is not sustainable, as our suburbs expand and we creep closer to what were once remote disposal places.

The problem is most apparent in water bodies, where our beaches and inland streams are paying the price. The pressure of these problems is now so great that authorities are moving to rectify the situation, but they can only move a fraction ahead of public opinion.

A project to recycle wastewater to save the Florida Everglades was rejected by the public because terms such as 'sewage', 'black water' and 'grey water' were used in the project description. Scientists world-wide took notice of the public's emotional overreaction, and now promote projects using terms such as 'recycling' and 'reuse'.

In 2002 the Premier of Victoria indicated that in future we will be treating sewage and recycling the water even for household purposes. The media overreacted because he was too far ahead of public opinion, but the idea is gaining ground.

Grey water

The average household produces 300–350 L of grey water daily. This amounts to 110 000–128 000 L/pa for the average household and represents a considerable potential resource which, if harnessed properly, could cater for a large part of garden watering.

The four main obstacles to community-wide use of grey water are:

- popular attitudes to odour, colour and turbidity (muddy appearance)
- public health concerns
- damage to the environment
- technology to refine water to acceptable purity.

Attitudes

We are all presented with the contradiction of the marketing sector urging us to use more chemicals in the home for a 'better lifestyle', while the regulatory sector says we should not release chemicals into the environment because of the damage they can do. Both sectors base their arguments on science. In both cases water is likely to be the transport medium.

Our attitudes are based to some extent on fact, but mostly on conditioning.

Grey water can smell and look horrible, but the use of a grease trap and simple filtration can overcome the most objectionable odours and turbidity.

Using only shower, bath and hand-basin grey water can reduce most of the problems because those sources do not involve many contaminants.

We have been conditioned, rightly, since our childhood to exercise care in hygiene. Wash our hands before meals, treat cuts with care to avoid infection – these are all sensible behaviour. This conditioning is reinforced by advertisements exhorting us to use chemicals to kill germs in just about every household situation.

There are three objections to excessive conditioning.

- We live with germs every day and our bodies are equipped to deal with most common germs. The most important countermeasures to infection are a healthy constitution and sensible hygiene. In any case, a sterile household is an impossible dream.
- The overuse of chemicals is producing supergerms by building up their resistance through indiscriminate application. This encourages industry to develop even more extreme cleaners that induce a further cycle of extra-supergerms. We only have to look at hospitals to see the evidence.
- Such chemicals are causing serious damage to our environment, and cost money both to purchase and to counteract.

The solution to the chemical problem in recycled water is in the hands of every householder. The choices we make at the supermarket and the use of cleansers in the home are ours. We can remove these deadly cocktails from our environment by choosing not to use them. It gives strength to the argument for recycling water at the household level rather than using the traditional method of bulk transport and treatment at the sewage plant.

Public health

While humans lived on farms or in small villages the disposal or reuse of waste was not a problem. Problems began as communities got larger, during the industrial revolution.

Sewerage and stormwater drainage systems were developed to get rid of serious diseases. Without them life in big cities would have been hazardous indeed. Public health was the driving force. This movement developed at a time when it was recognised that only government action could tackle community-wide problems and when society was also beginning to think industrially. But there were two major problems.

- Such systems require enormous capital outlay. Changing, maintaining and updating the technology involves further large capital outlay.
- Industrial wastes added to human waste in the system, further complicating reuse.

History has shown that changing or replacing highly capitalised and labour-intensive systems can be revolutionary and painful. In developing community-wide systems we have also developed structural inertia.

The main thrust of present wastewater research is targeted at the end point of the system, not the beginning. It has to deal with a range of chemical ingredients that our ancestors on the land never contemplated.

If the focus were shifted to the beginning of the system, the set of problems would change. Successful treatment would then involve simple changes in human behaviour, relatively simple waste products and the possibility for on-site reuse.

Damage to the environment

Coral reefs around the world are under major threat from excessive nutrients discharged from human settlements, including resorts built for tourists who want to enjoy the beauties of nature. Australia's inland rivers have become algae heaven and wildlife hell. The solutions for handling wastewater have just shifted the problems. One of the most effective ways of solving the problem is to develop low-cost and easily managed household water reuse systems.

Technology

There are several problems with existing wastewater reuse systems.

- Although the cost of the conversion component of the systems is within reach of some consumers, the cost and technical demands of subsurface garden distribution systems make them unattractive to most.
- Much of the household wastewater reuse technology on the market is produced by small entrepreneurs who may not have the capital or marketing capacity for large-scale production. The consequence is that such systems are relatively expensive and little used.
- Existing households already rely on the convenience of existing sewerage technology and developers of new housing tracts find it simpler to install conventional drainage and sewerage systems.
- Systems presently on the market do not meet the high water-quality standards set for total household reuse. Although they can be connected to bathroom outlets, toilet, kitchen and laundry outlets may not be connected. So not all the wastewater is reused.
- Treated wastewater can only be used in the garden by subsurface distribution; it can't be used even to flush toilets. There is only partial input and output to the systems – what is needed is total input/output reuse systems, like those that have been developed for astronauts.

The economic repercussions of the 'small is beautiful' approach to the problem is enormous. The fact that it involves a shift of reuse responsibilities from bureaucracies to individuals should not deter us. The technology is here now but it may need refinement and the economies of scale that mass production provides.

Total systems research

The US has pioneered wastewater reuse in modern times. In 1925, for example, the tourist facilities at the Grand Canyon recycled wastewater to flush toilets and water lawns. Most of the current methods of reuse have been driven by water shortages in the drier states, where research bodies are investigating reuse for recreational, agricultural and even domestic situations.

Some private citizens are developing methods of reuse. One interesting example is at a worm farm in Los Angeles where street runoff, mainly from overwatered lawns, was

Figure 12.1a: The hugely excessive gutter runoff from nearby suburban gardens is diverted into a worm farm water collection pit.

diverted into a holding pit where water hyacinth was used to absorb any solids and excess nutrients. The runoff was then pumped and sprayed onto compost and worm beds in a cocktail with nutrient-rich runoff from adjoining dairy farms. See Figures 12.1a, b and c.

Examples can also be found in South Africa, Europe, Japan and Israel, which recycles two-thirds of its effluent.

There is an interesting example of biofiltration in India, developed by Dr Uday Bhawalkar. The system uses an extended canna lily and reed bed incorporating earthworms to filter factory wastewater for reuse on a nearby garden. Even the slashed lilies and reeds are used as mulch. Some of the wastewater the system treats starts out as pretty horrible stuff, especially the effluent from soya sauce manufacture.

Figure 12.1b: The collection pit.

Figure 12.1c: The recycled water is used on worm beds.

Stormwater

Several recent housing estates in Australia have been harnessing stormwater on-site to create landscape features such as ponds and wildlife habitat.

Desalinisation

The process of reverse osmosis is the most used means of desalinisation of seawater and is used in some parts of the Middle East. The concept of solar evaporation desalinisation technology is currently being explored in Australia. The technique involves transporting seawater inland in a channel capped with a transparent canopy that uses the sun to evaporate the water and collect it after it has condensed. The salty water is then either returned to the sea or used as part of a salt extraction process. The project has the potential to change land settlement in Australia.

Appendix 1

The importance of water to plants

The whole plant

The healthy plant

Plant cells are made up of 80% or more water. Water is carried to the leaves by water-bearing tissue called the xylem. As the plant transpires water through leaf perforations called stomata, mostly on the underside of the leaves, it is replaced with water from the soil, via the roots. Many plants, such as geraniums, can also absorb water through their leaves. Plants take mineral nutrition from the soil along with the water.

The stressed plant

Without water, leaves curl and the plant cannot take up nutrients. Plants with few nutrients cease to grow and are prone to disease and insect attack.

If the soil dries out for an extended period the plant cannot take up moisture, but still loses it through the leaves.

The action of sun and wind help to dry out the plant. The plant's cells, which should contain water, shrink and the

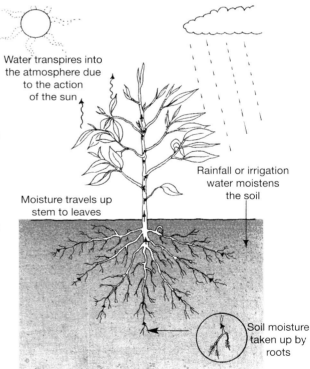

Figure A1.1: Water movement through the plant is initiated by the sun drawing water vapour from the leaves, starting the cycle which eventually draws water from the soil. The cycle is completed when rain or our watering moistens the soil.

leaves wilt. Growth ceases; the plant can die if the dryness is prolonged. Perhaps not so well known is the effect of frost, which dries the air near ground level. Frosts are more prevalent during drought. Plants suffer scorching on their leaves and even on stems.

If the soil is partially damp the plant roots will grow down in search of water.

The process of water uptake, transport through the plant, and transpiration is not fully understood, but each stage is important to the plant's survival. We will now take a more detailed look at the plant's water cycle.

Soil

Soil moisture occurs due to rainfall or irrigation. Water behaves differently in different types of soil.

The moisture is held in the soil as a film on and between the soil particles. It is held to the particles by a very strong force of interaction between soil molecules and water molecules.

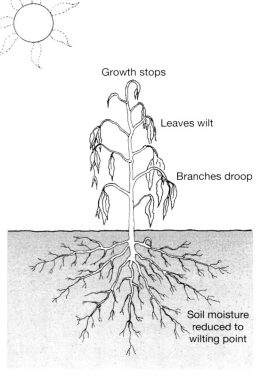

Figure A1.2: When soil moisture is insufficient the water cycle breaks down and the plant suffers stress, which can kill it if the dry spell continues.

When soil is saturated with water, it has reached 'field capacity'. If water continues to be applied beyond this stage, gravity causes it to drain away. Or if the soil is poor, it becomes waterlogged and muddy. At field capacity, the particles of soil are thickly coated in water. There is also moist air between the water-coated particles.

If the soil dries out to the point where it can dry no more under natural conditions, it has reached 'wilting point'. At wilting point, the coating of water is so thin and firmly held by the soil particles that plant roots cannot access it.

Table A1.1: Approximate transpiration rate (and water requirement) of plants on a summer day

Mature plant	Summer day transpiration (L)
Tomato	1
Sweetcorn	2
Sunflower	5
Apple tree	37
Palm tree	43
Willow	100+

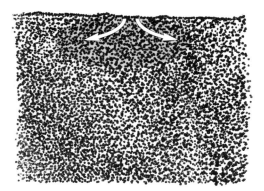

Figure A1.3a: In clay, water is slow-draining and tends to move sideways.

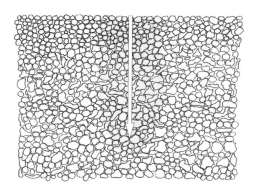

Figure A1.3b: In sand, water drains downwards relatively rapidly.

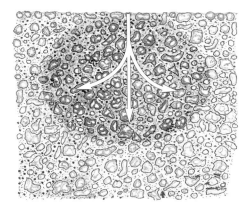

Figure A1.3c: In loamy soil that is rich in organic matter, water tends to move both sideways and downwards, and is retained longer to the benefit of plants.

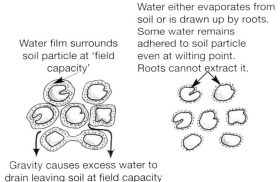

Figure A1.3d: Soil at field capacity and at wilting point.

Roots

The roots are in close contact with the soil particles and the fine root hairs draw in the soil moisture, which is absorbed through the root hair outer cuticle into the plant cells. The walls of each plant cell are permeable, so water continues to be drawn into and upwards through the plant.

Roots tend to grow toward the nearest source of available water.

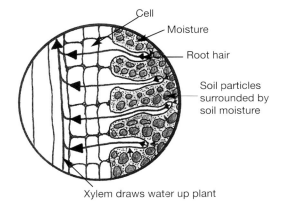

Figure A1.4: The xylem draws water up the plant from the soil, via the root hairs.

Stem

Within the stem are tube-like vessels called xylem through which water travels upwards, by a force involving capillary action, and is distributed to the leaves of the plant.

Leaves

The leaves require water for the process of photosynthesis, which produces food for the plant. Water enters the leaves by vessels called leaf veins. It then travels to and between cells where the photosynthesis occurs throughout daylight. Water not used in photosynthesis can evaporate through the leaf pores (stomata) into the outside atmosphere. In hot weather this evaporation can act as a cooling agent, just as it does in us when we perspire. That is why it is cooler in the shade of a tree on a hot day than under a non-living structure such as a verandah.

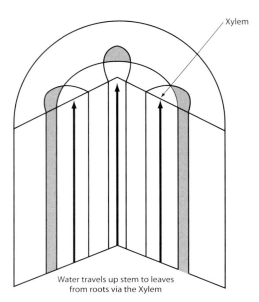

Figure A1.5: An enlarged section of the stem showing water being drawn up the plant from the roots, through the xylem.

Table A1.2: Approximate number of stomata per square cm in some vegetables

Sweetcorn	6 000
Tomato	7 000
Potato	10 000
Cabbage	18 000

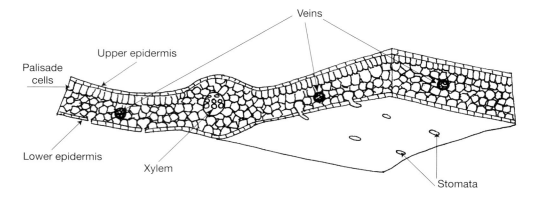

Figure A1.6: Cross-section of the leaf. Water arrives from the stem via the xylem and enters the leaf cells. In the warmth of daytime it evaporates through tiny opening valves called stomata.

Flowers and fruit

The flowers and fruit are the reproductive organs of the plant, essential for the continuation of the species. The following is a very simplified account of a process that varies among many species of flowering plants.

Sepals and petals protect the developing stamens and carpels. Pollination occurs after the pollen is transported to the stigma by wind or insects. The pollen grain sends a tube down to the ovary, where germination takes place and the seed forms. The ovary and the seed together form the fruit.

The cells of the fruit and flowers are similar to those of the rest of the plant, and suffer the same stress when water is in short supply. As a result, fruit may not form or it may shrivel and lack flavour and nutrition. Such drought-induced deficiencies apply whether the fruit is an apple, pea, pear, plum, tomato or pumpkin.

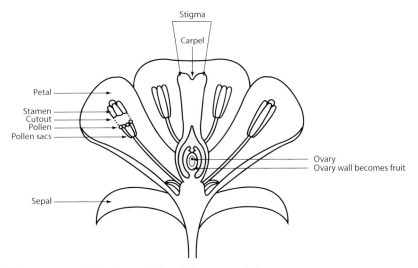

Figure A1.7: The structure of the flower before it becomes fruit.

Plant cells

The plant is made up entirely of cells, which are not visible to the naked eye. They comprise a vacuole, which is filled with water, surrounded by cytoplasm, which holds the cell nucleus, and an enclosing permeable cell wall. The water in the vacuole applies outward pressure on the cell wall, giving the cell and hence the plant its structural strength.

Plant stress occurs when the moisture level of the soil around the roots falls below field capacity and the root hairs can no longer draw in the moisture. This sets up a chain reaction of moisture deficiency in the plant, as the sun continues to draw water from the leaves.

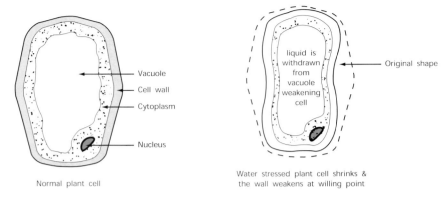

Figure A1.8: The healthy cell on the left has a strong structure because it holds adequate water. The stressed cell has insufficient water to maintain a strong cell wall, which starts to collapse as the plant reaches wilting point. As a result, the plant can be attacked by disease and pests.

This affects every living cell within the plant. The loss of moisture causes cells to shrink and shrivel, which in turn causes a loss of structural strength throughout the plant. As these building-blocks shrink so the plant starts to wilt all over.

Heat-stress wilting occurs more rapidly during hot weather. Unless the soil moisture is replaced rapidly the plant will suffer so much stress it may die.

Fatal moisture stress can occur over cyclic periods as short as a day or as long as several years, depending upon the structural design of the species. Plant groups with longer cycles are of particular interest to the waterwise gardener.

Plant strategies to overcome dry conditions

All plants die without water, but those that have evolved in dry areas have developed effective survival strategies. These survival mechanisms can occur at the root, stem or leaf stage of the transpiration process.

Roots

The root's survival strategy is to grow deeper as water recedes from the surface in dry times. In deep soil, it is surprising how deep roots will go in search of moisture. Some native plant roots will go very deep to tap into the watertable and will never need extra watering.

When selecting plants ask your nursery supplier about deep-rooting species. Make sure you keep such species well away from underground pipes, as they tend to find small cracks or strangle pipes in searching for water.

Some books recommend keeping these plants at least 2 m away from pipes. It has been my experience that the roots of vigorous plants, including large eucalypts and acacias, poplars and willows, travel up to 20 m in loose soils. Most gardens are too small for these larger trees anyway. For a comprehensive list of drain stranglers and cloggers see Appendix 5.

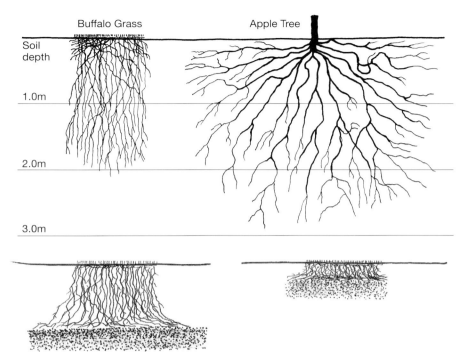

Figure A1.9: Root systems in deep soil (top) and in shallow soils (bottom). The roots in deep soil penetrate to great depths, giving the plant a greater survival chance in dry conditions. The roots in shallow soil with a hardpan of clay near the surface are restricted in their ability to survive dry conditions. The lawn roots in very shallow soil will eventually choke on their own dense rootmat, which becomes an impenetrable tangle for water and nutrients.

Stem

Some plants can store massive amounts of reserve moisture in their stems during wet times to sustain them through extended dry periods.

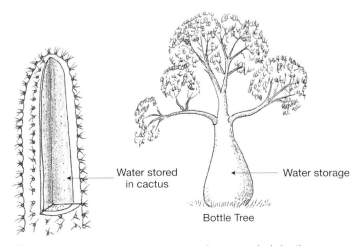

Figure A1.10: Stems used as water storage to survive extended dry times.

Leaves

There are several leaf survival strategies.

- Some plants store water in fleshy leaves.
- Plants such as cacti have reduced the size of their leaves to spikes, to reduce evapotranspiration.
- Some plants have a tough, thick, less-permeable cuticle.
- The stomata may be deeply embedded below the surface.
- The stomata may be less numerous.
- Leaves may hang vertically so they are less exposed to the sun's rays. Some plants can actually change leaf angle from horizontal to vertical in dry summer.

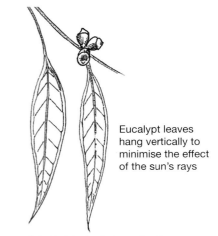

Eucalypt leaves hang vertically to minimise the effect of the sun's rays

Figure A1.11: Vertically hanging leaves.

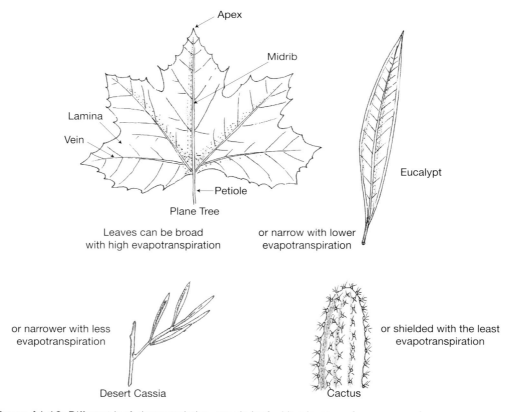

Figure A1.12: Different leaf characteristics: maple leaf with a large surface area and numerous stomata is subject to high evapotranspiration water loss; the tough gumleaf has many fewer stomata; the desert cassia leaf has an even smaller surface area; the cactus with spikes for leaves offers minimal surface area to the sun.

- Leaves may be hairy to provide themselves with shade and windbreaks to reduce evapotranspiration.
- Many leaves are reflective, with lighter colouring or shiny reflective surfaces. Some even gather salt on their surface to enhance their reflectivity.

Overall plant strategies

There are many plant survival strategies.

- In dry times plants slow down their metabolic rates. Some may appear to be dead, but life remains in a sheltered core. These plants burst into life after rain. Such plants may suffer fungal diseases if overwatered in the home garden.
- Many can extract moisture from the predawn air.
- Plants tend to produce more seed, as a species survival strategy, when stressed by a water shortage. The plant may die but with its last heroic gasp it produces seed for the next generation. Many of these seeds have tough cuticles that allow them to lie dormant under the most hostile conditions then germinate with the first rains. Some such as wattle germinate best after fire.
- Many plants self-mulch by dropping their leaves to form an insulating layer on the soil surface.
- Water-efficient plants may have a strong structure to prevent damage due to wilting.
- Some plants can influence the soil nutrients around their roots to improve moisture retention. Soil organic matter can hold up to five times its weight in

Figure A1.13: A casuarina tree both shades its roots and self-mulches.

moisture. While this may not be considered a plant strategy, many add organic matter to the soil by various means.

- The roots, including the nodulised roots of legumes, remain in the soil after the plants die, thus adding organic matter to the soil.
- Root parts, including the root hairs, die and are replaced, adding to soil organic matter.
- Leaf litter supports a range of small organisms which add their wastes and corpses to soil organic matter.
- Smaller plants use the shade, reduced wind velocity and humidity created by larger plants to reduce moisture loss.

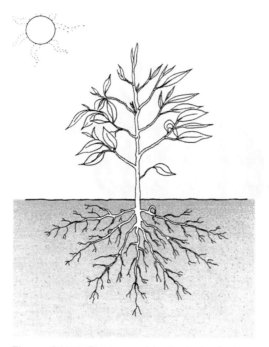

Figure A1.14: Plants provide shade for their own roots.

Figure A1.15: Taller hardier plants provide shade for smaller companion plants.

Appendix 2

Australian plants tolerating very dry conditions

Botanical names are shown first because they are consistent in meaning, whereas common names vary from district to district, and even within districts. The state of origin is also given, to indicate where the plant grows naturally and which latitudes it does best in.

These plants are tough enough to last for six months without water. They will require good drainage, preferably with sandy soil mounding.

Ground covers	Common name	Comments (including native locations)
Acacia	Wattle	
acanthoclada	Harrow wattle	NSW, Vic., SA, WA. Partial or full sun. Spiky.
cometes		WA. Partial or full sun.
dictyophleba		Qld, WA, SA, NT. Sandy soil.
menzelii		SA. Partial or full sun. Sticky leaves.
merrallii		SA, WA. Clay. Semi-shade to full sun.
oxyclada		WA. Well-drained clay to sandy soil, semi-shade to sun.
spinescens	Spiny wattle	NSW, Vic., SA. Sandy to light clay, semi-shade to full sun.
Atriplex		
rhagodioides	Silver saltbush	NSW, Vic., SA. Various soils, partial or full sun, salt-tolerant, frost-hardy, fire-retardant.
Dampiera		
lavandulacea	Lavender dampiera	WA. Attractive semi-shade to sun.
Disphyma		
australe	Noon flower	Qld, NSW, Vic., Tas., SA, WA. Semi-shade to full sun, salt and frost-tolerant, fire-retardant.
blackii	Noon flower	Qld, NSW, Vic., Tas., SA, WA. Semi-shade to full sun, salt and frost-tolerant, fire-retardant.

Enchylaena		
tomentosa	Ruby saltbush	Qld, NSW, Vic., SA, WA, NT. Alkaline soils, partial to full sun, salt-tolerant, berries.
Eremophila		
crassifolia	Trim emu bush	Vic., SA. Well-drained soil, full sun. Blue flower.
densifolia		WA. Semi-shade to sun. Purple flower.
glabra (prostrate)	Common emu bush	Qld, NSW, Vic., SA, WA, NT. Well-drained, semi-shade to sun. Yellow-orange flower.
ionantha	Violet flowered eremophila	WA. Most soils, full sun.
macdonnellii		Qld, NSW, SA, NT. Most drained soils, full sun.
maculata	Native fuchsia	Qld, NSW, Vic., SA, WA, NT. Most soils, semi-shade to sun.
Eutaxia	Eutaxia	
microphylla		Qld, NSW, Vic., Tas., SA, WA. Most soils, full shade to sun.
Frankenia	Southern sea heath	
pauciflora		WA. Most soils, full sun. Some forms in SA and Vic.
Glischrocaryon		
aurea	Common popular flower	SA, WA. Sandy soil, full sun.
behrii	Golden pennants	NSW, Vic., SA. Most soils, partial shade to full sun.
Glycine		
tabacina	Vanilla glycine	Qld, NSW, Vic., SA, WA. Dappled light, climber.
Gossypium		
sturtianum	Sturt's desert rose	Qld, NSW, SA, WA, Vic., NT. Sandy soil, full sun.
Grevillea		
aspera		SA. Most soils, semi-shade to full sun.
disjuncta		WA. Most soils, semi-shade to full sun.
ilicifolia		NSW, Vic., SA. Sandy soil, semi-shade to full sun.
rogersil		SA. Most soils, semi-shade to full sun.
steiglitziana	Brisbane Ranges grevillea	Vic. Well-drained soil. Semi-shade.
Helichrysum		
acuminatum	Alpine everlasting	NSW, Vic., Tas. Most soils, sun.
apiculatum	Yellow buttons	Australia. Medium to heavy soil, sun.
baxteri	Fringed everlasting	NSW, Vic., Tas., SA. Light soil, sun
braceatum	Immortelle	Australia. Most soils, sun.
diosmifolium		Qld, NSW. Light to medium soil, well-drained, sun.
hookeri	Alpine everlasting	NSW, Vic., Tas. Well-drained, sun, cool climate.
semipapposum	Clustered everlasting	Qld, NSW, Vic., Tas., SA. Light to heavy soil, sun.
thyrsoideum	Cascade everlasting	NSW, Vic., Tas. Well-drained, sun.
Helipterum		
albicans	Hoary sunray	Qld, NSW, Vic. Good drainage, sun.
roseum	Rosy sunray, pink everlasting	WA. Sandy to loam soil, full sun.

Appendix 2 | 137

Hibbertia		
many forms	Guinea flower	
Hymenosporum		
australis	Australian indigo	Australia. Light to medium soil, sun.
flavum	Native frangipani	Qld, NSW. Light to medium soil, sun, frost-tender.
indegofera		
Jasminum		
linearis	Desert sasmine	Central Australia. Vine, light to heavy soil, shade.
Kunzea		
pomifera	Muntries	SA, Vic. Sandy soil, full sun. Edible berries.
Lasiopetalum		
schulzenii	Velvet bush	SA, Vic. Sandy soil, sun.
Leucopogon		
biflorus	Twin flower, beard heath	Qld, NSW, Vic. Sandy soil, semi-shade.
Myoporum		
debile	Amulla	NSW, Qld. Well-drained, sun.
parvifolium	Creeping boobiala	SA, Tas., Vic. WA. Well-drained, full sun, fire-retardant.
Phebalium		
bullaium		SA, Vic. Well-drained, sun.
glandulosum		NSW, Qld, SA, Vic. Well-drained, sun.
Prostanthera		
aspalathoides	Scarlet mintbush	NSW, SA, Vic. Well-drained, sun.
Ptilotus		
erubescens	Pink mulla mulla	Inland Australia. Well-drained, sun.
exaltatus	Tall pussy tails	Widespread. Well-drained, sun, frost-tender.
Rhagodia		
baccata	Coastal saltbush	NSW, Vic., Tas., SA, WA. Light soil, sun.
nutans	Saltbush	Widespread. Light to medium soil, sun.
Swainsona		
canescens	Grey pea	WA. Light soil, sun.
greyana		SA, Vic., NSW, Qld. Heavy soil, sun.

Small to medium shrubs	Common name	Comments (including native locations)
Acacia	Wattles	These shrubs are relatively short-lived.
aspera	Rough wattle	Qld, NSW, Vic., Tas., SA. Semi-shade, drainage.
assimilis		WA. Well drained, sun.
conferta	Golden top	Qld, NSW. Various soils, semi-shade to full sun.
farinosa	Mealy mallee	NSW, Vic., SA. Light to medium soils, semi-shade to sun.
hakeoides	Hakea wattle	Qld, NSW, Vic., SA, WA. Most soils, semi-shade to full sun. Windbreak.
ligulata	Small cooba	NSW, Vic., SA, WA, NT. Sandy soil, full sun. Windbreak.
loxophylla		
var.nervosa		WA. Light to medium soil, semi-shade to sun.

myrtifolia	Myrtle wattle	All states. All drained soils, semi-shade to full sun.
rigens	Nealie	Qld, NSW, Vic., SA. All drained soils, semi-shade to sun.
spathulata	Acacia	WA. All drained soils, full sun, frost-tender.
steedmanii		WA. Well-drained soils, semi-shade to sun.
tetragonaphylla	Kurana	Qld, NSW, SA, WA, NT. Well-drained soils, full sun.
williamsonii	Whirrakee	Vic. Most soils. Semi-shade to full sun.
Atriplex		
nummularia	Old man saltbush	Inland Australia. Most soils, full sun, salt-tolerant, fire-retardant. Windbreak.
Cassia		
nemophila	Desert cassia	NSW, Vic., SA, WA, NT. Most soils, semi-shade to sun.
sturtii		NSW, SA, WA, NT. Well-drained, sun.
Eremophila		
calorhabdos		Inland Australia. Well-drained, full sun.
denticulata		SA, WA. Well-drained, full sun.
drummondii		WA. Most soils, sun.
freelingii		Inland Australia. Well-drained, sun.
glabra	Tarbush	Inland Australia. Well-drained, sun.
laanii		WA. Well-drained, sun.
latrobei		SA, NSW, Qld, NT, WA. Well-drained, sun.
longifolia	Berrigan	All except NT. Well-drained, light to medium soil, sun.
maculata	Spotted emu bush	Inland Australia. Medium to heavy soils, sun.
polyclada	Twiggy emu bush	NSW, Qld, SA, Vic. Well-drained, light to medium soil, sun.
serrulata		NSW, SA, WA, NT. Well-drained, sun.
Grevillea		
inconspicua		WA. Well-drained, sun. Not showy.
rosmarinifolia	Rosemary grevillea	NSW, Vic. All soils, sun. Windbreak.
Hakea		
circumalata		WA. Sandy soil, sun.
costata		WA. Well-drained, sun.
Lasiopetalum		
baueri	Velvet bush	NSW, Vic., SA. Light to medium soil, sun.
behri	Pink velvet bush	NSW, Vic., SA. Light to medium soil, sun.
ferrugineum	Velvet bush	NSW, Vic. Well-drained, light soil, sun.
Myoporum		
desertii	Turkey bush	Qld, NSW, Vic., SA, WA. Light to heavy soil, sun.
insulare	Boobialla	NSW, Vic., SA, WA. Most soils, sun, fire-resistant. Windbreak.
Prostanthera		
behriana	Mintbush	SA. Well-drained, sun.
Westringia		
eremicola		Qld, NSW, Vic. Light to medium soil, sun.
fruticosa	Coastal rosemary	Qld, NSW. Light soils, sun.
fruitcosa x glabra		Qld, NSW, Vic. Light soils, sun.

Larger shrubs and trees	Common name	Comments (including native locations)
Acacia		
acuminata	Raspberry jam wattle	WA. Light soil, well-drained, sun.
amoena	Boomerang wattle	Qld, NSW, Vic. Light soil, well-drained, sun.
aneura	Mulga	All states. Light soil, well-drained, sun.
argyrophylla	Silver mulga	SA. Light to heavy soil, well-drained, sun. Warmer areas.
arida		WA, NT. Well-drained, sun. Suits north-west.
beckleri	Acacia	NSW, Vic. Light soil, well-drained, sun.
brachystachya	Umbrella mulga	Inland Australia. Well-drained, sun.
calamifolia	Wallowa	Qld, NSW, SA. Light to medium shallow soils, sun.
cambagei	Gidgee	Qld, NSW, NT. Medium to heavy soils, sun. Windbreak.
cardiophylla	Wyalong wattle	NSW. Medium soils, sun.
colletioides	Wait-a-while	NSW, Vic., SA, WA. Medium to heavy soil, sun.
cunninghamii		Qld, NSW. Well-drained, sun, coastal.
farnesiana	Perfumed wattle	NSW, SA, NT, WA, Qld. Most soils.
gladiiformis	Swordleaf wattle	NSW. Light soils, sun.
harpophylla	Brigalow	NSW, Qld. Medium to heavy soil, sun.
iteaphylla	Willow leaf, Gawler	SA. Light to medium soil, sun.
ligulata (form)	Small cooba	Inland Australia. Medium to heavy soil, sun.
oswaldii	Umbrella wattle	All mainland Australia. Light to medium soil, sun.
pendula	Weeping myall	Qld, NSW, Vic. Medium to heavy soil, sun.
peuce	Acacia	Qld, SA, NT. Most soils, sun.
pycnantha	Golden wattle	NSW, Vic., SA. Most shallow soils, sun.
quornensis		SA. Light to heavy soils, semi-shade to sun. Windbreak.
retinodes	Silver wattle, wirilda	Vic., Tas., SA. Light to medium soil, sun.
rossei	Ross's wattle	WA. Light, drained soil, sun.
sowdenii	Western myall	SA. Medium to heavy soil, sun.
spectabilis	Eumung	Qld, NSW. Medium to heavy soil, sun.
stenophylla	Native willow	All mainland Australia. Heavy soils, sheltered position.
terminalis	Cedar wattle	Qld, NSW, Vic. Light to medium soils, sun.
uncinata	Weeping wattle	Qld, NSW. Most well-drained soils, long-flowering understorey.
victoriae	Bramble wattle	Qld, NSW, Vic., SA, NT, WA. Medium to heavy soils, sun.
Actinostrobus		
pyramidalis	Swan River cypress	WA. Most well-drained soils, sun, salt-tolerant.
Adansonia		
gregorii	Bottle tree	Inland Australia. Most soils, sun, frost-tender.
Albizia		
lebbeck	White siris	Qld. Most drained soils, sun, shade tree, scented, frost-tender.

Alyogyne		
hakeifolia	Desert rose	SA, WA. Light to medium soil, sun. Trumpet flowers.
Brachychiton		
gregorii	Desert kurrajong	SA, WA, NT. Light to medium soil, full sun. Tough.
populneus	Kurrajong	Qld, NSW, Vic., NT. Most soils, sun, frost tender.
rupestris	Bottle tree	Qld, NSW. Most soils, sun.
Callistemon		
lilacinus	Lilac bottlebrush	NSW. Well-drained soils, sun.
Callitris		
columellaris	Murray pine	Qld, NSW. Light to medium soil, sun.
drummondii	Cypress pine	WA. Most soils, sun.
endlicheri	Black cypress pine	Qld, NSW, Vic. Sandy soils, sun. Long-lived.
preissii	Slender cypress pine	Qld, NSW, Vic., SA, WA. Sandy, sun, shade-tree, salt-tolerant.
roei	Cypress pine	WA. Sandy soil, sun, rare.
Cassia		
artemisioides	Silver cassia	NSW, SA, NT. Well-drained soil, semi-shade to sun.
brewsteri	Leichhardt bean	Qld. Well-drained, sun, frost-tender.
pleurocarpa		Qld, WA, NT. Drained, sun, shelter from wind.
Casuarina		
campestris		WA. Well-drained, sun.
cristata	Belah	NSW, Vic., SA, WA. Light to heavy soils, sun.
decaisneana	Desert oak	Qld, WA, NT. Light to medium soil, sun.
luehmanniana	Bull oak	Qld, NSW, Vic., SA. Various soils, sun.
Cienfugosia		
patersonii	Sturt's desert rose	Qld, NSW, SA. Most soils, sun. Hibiscus-like flowers.
Crotalaria		
cunninghamii	Parrot plant	Central Australia. Light soil, drained, sun.
Cycas		
revoluta	Fossil cycad	Evergreen fern, poor soil, sun, wind-prone.
Eremophila		
alternifolia	Emu bush	NSW, SA. Light to medium alkaline soil, sun.
longifolia	Berrigan	All except NT. Light to medium soil, sun.
maculata	Spotted emu bush	Inland Australia. Medium to heavy soil, sun.
Eucalyptus	Use mallees for suburban gardens	
albida	White-leafed mallee	WA. Sandy soil, well-drained, sun.
angulosa	Ridge-fruited Mallee	NSW, Vic., SA, WA. Light soil, sun, frost-tender.
behriana	Broad-leafed mallee box	NSW, Vic., SA. Light to medium soil, sun.
brachycorys	Cowcowing mallee	WA. Light alkaline soil, sun.
brachyphylla	Short-leafed mallee	WA. Well-drained granite soil, sun.
burdettiana	Burdetts gum	WA. Well-drained, sun, frost-tender.
burracoppinensis	Burracoppin mallee	WA. Sand, well-drained, sun, frost-tender.
caesia	Gungurru	WA. Well-drained, sun, frost-tender, needs wind shelter.

calycogona	Gooseberry mallee	WA. Well-drained, sun, frost-tender.
campaspe	Silver-topped gimlet	WA. Sandy loam, sun.
carnea	Carnes blackbutt	WA. Clayey loam, sun.
cladocalyx	Sugar gum	SA. Most soils, sun, try small form in home garden.
concinna	Victoria Desert mallee	WA. Light sandy soil, sun.
cornuta	Yate	WA. Loamy soil, sun.
coronata	Crowned mallee	WA. Well-drained, sun.
crucis	Silver mallee	WA. Light soil, well-drained, sun.
curtisii	Plunkett mallee	Qld. Sandy soil, well-drained, sun.
cylindriflora	White mallee	WA. Sandy loam, sun.
dealbata	Tumbledown gum	NSW. Shallow poor soils, sun.
desmondensis	Desmond mallee	WA. Light medium soil, sun, frost-tender, open foliage.
dielsii	Cap-fruited mallee	WA. Heavy soil, well-drained, sun.
diptera	False gimlet	WA. Heavy alkaline soil, sun.
diversifolia	Soap mallee	SA. Sandy alkaline soil, sun.
dumosa	Congoo mallee	NSW, SA, Vic. Light to medium soil, well-drained, sun.
dwyeri	Dwyers mallee gum	NSW. Shallow poor soil, sun.
ebbanoensis	Sandplain mallee	WA. Sandy soil, well-drained, sun.
eremicola	Nawa mallee	SA. Well-drained, sun.
eremophila	Tall sand mallee	WA. Light to heavy soils, sun.
erythrandra	Rosebud gum	WA. Light soil, well-drained, sun, sheltered position.
erythrocorys	Illyarrie	WA. Medium to heavy soils, sun, frost-tender.
erythronema	Red-flowered mallee	WA. Light to heavy soils, sun.
eudesmoides	Myallie	WA. Sandy soil, well-drained, sun.
foecunda	Narrow-leafed red mallee	WA. Alkaline sands, sun, frost-tender. Windbreak.
forrestiana	Fuchsia mallee	WA. Light to heavy soil, well-drained, sun.
froggattii	Kamarooka mallee	Vic. Medium to heavy soil, sun.
gamophylla	Twin-leafed gum	Central Australia. Deep sands, sun.
gardneri	Blue mallee	WA. Light to medium sand, sun.
gillii	Curly mallee	SA. Light to medium acid soil, sun.
gomphocephala	Tuart	WA. Light alkaline soils, sun, salt-tolerant.
goniantha	Jerdacottup mallee	WA. Light to medium soil, well-drained, sun.
gracilis	White mallee	Vic. Light soil, well-drained, semi-shade, fire-retardant.
grossa	Phillips River gum	WA. Clay loams, sun, likes moist site.
incrassata	Ridge-fruited mallee	SA. Light soil, well-drained, sun.
intertexta	Inland red box	Inland Australia. Light to heavy soils, sun.
jutsonii	Jutsons mallee	WA. Sand and sandy loam, sun.
kingsmilli	Kingsmills mallee	WA. Sandy acid soil, sun.
kruseana	Bookleaf mallee	WA. Granite sands, sun.
lansdowneana	Crimson mallee box	SA. Light to medium soil, well-drained, sun, wind-prone.
lehmannii	Bushy yate	WA. Light to medium well-drained soil, sun.
leucoxylon	Yellow gum	Vic. Clay soil, sun.
leucoxylon rosea	Red-flowered yellow gum	Vic. Clay soil, sun

macrandra	Long-flowered marlock	WA. Light to heavy soil, sun.
macrocarpa	Mottlecah	WA. Sandy soil, well-drained, sun, frost-tender.
mannensis	Mann Ranges mallee	SA, WA, NT. Light soil, drained, sun.
microcarpa	Grey box	Qld, NSW, Vic., SA. Medium to heavy soils, sun, can take poor drainage.
morrisbyi	Morrisbys bum	Tas. Well-drained, sun.
morrisii	Grey mallee	Qld, NSW, Central Australia. Light to medium soil, well-drained, sun.
normantonensis	Normanton box	Qld. Shallow poor soil, sun.
nova-angelica	New England blackbutt	NSW. Light to medium soil, well-drained, sun.
nutans	Red-flowered moort	WA. Sandy loam, sun, frost-tender.
oleosa	Giant mallee	Inland Australia. Sandy loam, well-drained, sun.
orbifolia	Round-leafed mallee	WA. Light to medium sandy soil, sun.
ovularis	Small-fruited mallee	WA. Clayey soils, sun.
pachyphylla	Red-bud mallee	Central Australia. Light soil, well-drained, sun.
platypus	Round-leafed moort	WA. Light to medium soil, poorly drained, sun decoration.
platypus var. heterophylla		WA. Alkaline sand, sun.
polyanthemos	Red box	NSW, Vic. Poor shallow soil, drained, sun.
polybractea	Blue-leafed mallee	NSW, Vic. Acid soil, well-drained, sun. Windbreak.
polycarpa	Red bloodwood	Qld, NSW, WA, NT. Most well-drained soils, sun, tropical.
populnea	Poplar box	Qld, NSW. Heavy soils, sun, shade tree.
preissiana	Bell mallee	WA. Acid loam, sun, frost-tender. Brilliant flower. Subtropical.
pyriformis	Pear-fruited mallee	WA. Sandy loam, sun.
redunca	Black marlock	WA. Light to medium soil, sun.
salmonophloia	Salmon gum	WA. Light to medium soil, well-drained, sun.
salubris	Fluted gum	WA. Medium to heavy soils, sun.
sargentii	Salt River gum	WA. Light to medium soil, sun, salt-tolerant.
scoparia	Wallangarra white gum	Qld, NSW. Light to medium soil, well-drained, sun.
sideroxylon	Ironbark	NSW, Vic. Light to medium soil, well-drained, sun.
sideroxylon rosea	Red-flowered ironbark	NSW, Vic. Light to medium soil, well-drained, sun.
similis	Inland yellow jacket	Qld. Sandy or loamy soils, sun.
socialis	Grey mallee	Inland Australia. Medium to heavy soils, sun.
spathulata	Swamp mallee	WA. Heavy clay loam, sun.
steedmanii	Steedman's mallee	WA. Light soil, sun.
stoatei	Scarlet pear gum	WA. Light to medium soil, well-drained, sun, frost-tender.
stowardii	Fluted horn mallee	WA. Light to heavy soil, sun, frost-tender.
stricklandii	Yellow-flowering blackbutt	WA. Sandy loam soils, sun.
tetragona	Silver marlock	WA. Light well-drained soil, sun, frost-tender.
tetraptera	Square-fruited mallee	WA. Poor sandy soil, sun.

todtiana	Prickly bark	WA. Light soil, well-drained, sun, frost-tender.
torquata	Coral gum	WA. Well-drained, sun.
'Torwood'	Torwood	WA. Well-drained, sun, best of torquata and woodwardii.
transcontinentalis	Boongul	WA. Most soils, sun.
uncinata	Hook-leafed mallee	WA. Light to medium sandy soil, sun.
viridis	Whipstick mallee	Qld, NSW, Vic., SA. Light to medium soil, well-drained, sun.
websterana	Websters mallee	WA. Light to heavy granite soils, sun.
woodwardii	Lemon-flowered gum	WA. Light to heavy soil, well-drained, sun.
yalatensis	Yalata mallee	SA, WA. Light shallow alkaline soil, sun.
Flindersia		
maculosa	Leopard tree	Qld, NSW. Medium to heavy soil, sun.
Geijera		
parviflora	Wilga	Qld, NSW, Vic., SA, WA. Light to heavy soils, sun.
Grevillea		
argyrophylla	Silvery-leafed grevillea	WA. Most drained soils, likes alkaline, sun.
juncifolia	Honeysuckle grevillea	Qld, NSW, SA, WA, NT. Sandy soil, sun.
petrophiloides	Pink pokers	WA. Light soil, well-drained, sun.
pterosperma	Desert grevillea	NSW, Vic., SA, WA, NT. Light soil, well-drained, sun.
robusta	Silky oak	Qld, NSW. Medium to heavy soil, sun, frost-tender.
rosmarinifolia (form)	Rosemary grevillea	NSW, Vic. Light to medium soil, sun.
striata	Beefwood	Qld, NSW, SA, NT. Heavy soils, sun.
Hakea		
divaricata	Corkbark	SA, WA, NT. Well-drained, sun.
Melaleuca		
armillaris	Bracelet honey myrtle	NSW, Vic., Tas. Most soils, sun.
bracteata	White cloud tree	Qld, NSW, SA, WA, NT. Medium to heavy soil, sun.
decussata	Totem poles	Vic., SA. Most well-drained soils, sun.
elliptica	Granite honey myrtle	WA. Well-drained, sun.
huegelii	Chenille honey myrtle	WA. Light to medium soil, well-drained, sun.
hypericifolia	Red-flowering paperbark	Qld, NSW. Light to medium soil, sun.
lanceolata	Moonah	Mainland Australia. Light to medium soil, sun.
Melia		
azedarach	White cedar	Qld, NSW, WA, NT. Most soils, sun to semi-shade, fruit poisonous.
Myoporum		
montanum	Western boobialla	Qld, NSW, Vic., SA, WA, NT. Most soils, sun.
Pandorea		
doratoxylon	Spearwood Bush	SA. Light well-drained soil, sun.
Pittosporum		
phillyreoides	Weeping pittosporum	Inland Australia. Most soils, sun.
Westringia		
longifolia		Qld, NSW. Light to medium soil, well-drained, sun.

Xanthorrhoea		
arborea	Grass tree	Inland Australia. 2 m, well-drained, sun, frost-tender.
australis	Grass tree	Vic., Tas., SA. 1.5 m, well-drained, sun, frost-tender.
macronema	Grass tree	Inland Australia. 1 m, well-drained, sun, frost-tender.
preissii	Grass tree	WA. 5 m, well-drained, sun, frost-tender.

Appendix 3

Exotic drought-tolerant plants

Trees, shrubs and herbs

Species	Common name	Comments
Abies		
borissii-regis	Balkam fir	Evergreen, 30 m, deep acid soil, sun or shade.
bracteata	Fringed spruce	Evergreen, 50 m, deep alkaline soil, sun or shade.
cephalonica	Grecian fir	Evergreen, 35 m, deep alkaline soil, sun, likes cool.
concolor	Colorado white fir	Evergreen, 45 m, most soils, sun or shade.
grandis	Giant fir	Evergreen, 75 m, most soils, sun or shade.
homolepis	Nikko fir	Evergreen, 25 m, most soils, sun or shade.
koreana	Korean fir	Evergreen, 15 m, most soils, sun or shade.
lasiocarpa	Alpine fir	Evergreen, 40 m, acid soils, sun or shade.
magnifica	Red fir	Evergreen, 60 m, deep soils, sun or shade.
nordmanniana	Caucasian fir	Evergreen, 60 m, deep soils, sun.
numidica	Algerian fir	Evergreen, 30 m, most soils, sun or shade.
pindrow	Himalayan fir	Evergreen, 60 m, most soils, sun or shade.
pinsapo	Spanish fir	Evergreen, 30 m, most soils, sun or shade.
Agapanthus		
africanus	Agapanthus	Perennial, 1 m, most soils, sun.
orientalis	Agapanthus	Perennial, 1 m, most soils, sun.
Agave	Various	Evergreen, spiky, can spread.
Ailanthus		
altissima	Tree of Heaven	Deciduous, 25 m, most soils.
Albizia		
julibrissin	Silk tree	Deciduous, 10 m, well-drained.
Aloe	Various	Evergreen, spiky, can spread.

Anagyris		
foetida	Stinkwood	Small tree
Aptenia		
cordifolia	Noon flower	Evergreen vine, most drained soils, sun, frost-tender.
lancifolia	Noon flower	Evergreen vine, most drained soils, sun, frost-tender.
Arbutus		
andrachne	Asian strawberry tree	Evergreen, 6 m, well-drained, sun.
Arctostaphylos		
columbiana	Hairy mananita	Evergreen, 3 m, well-drained, sun.
manzanita	Manzanita	Evergreen, 4 m, most soils, sun or shade.
uva ursi	Bearberry	Evergreen, 0.2 m, most soils, sun or shade.
Argemone		
ochroleuca	Mexican poppy	Annual, 1 m, most soils, sun.
Artemisia		
campestris	Field southernwood	Perennial, 1 m, well-drained, sun.
cini	Levant wormwood	Perennial, 1 m, well-drained, sun.
dracunculus	Tarragon	Perennial, well-drained, sun.
ludoviciana		Evergreen, 1 m, well-drained, sun.
nutans	Silver wormwood	Evergreen, 1.5 m, well-drained, sun.
pontica	Roman wormwood	Perennial, 1.5 m, most soils, sun.
tridentata	Sage bush	Evergreen, 2 m, well-drained, sun.
vulgaris	Mugwort	Perennial, 1.2 m, most soils, sun or shade.
Aspidistra		
elatior	Aspidistra	Perennial, 1 m, sandy peat, shade.
Atriplex		
canescens	Four-winged saltbush	Evergreen, 2 m, well-drained, sun.
hamiloides		Evergreen, 2 m, sand, sun, medicinal
hastata		Perennial, 1 m, salty soils, sun.
Azadirachta		
indica	Neem	Evergreen, 5 m, well-drained, sun, medicinal.
Baccharis		
pilularis	Coyote brush	Ground creeper, well-drained, sun.
sarothroides	Desert broom	Evergreen, 2 m, well-drained, sun.
Ballota		
nigra	Black horehound	Perennial, 1 m, well-drained, sun, medicinal.
Baptista		
tincoria	Yellow broom	Perennial, 1 m, well-drained soil, sun.
Bauhina		
hookeri frost-tender.	Mountain ebony	Evergreen, 15 m, light to medium soils, sun,
Bergenia		
crassifolia	Pig squeak	Evergreen, 0.5 m, well-drained, sun.

Bignonia		
catalpa	Tree bignonia	Deciduous, 20 m, most soils, sun or shade, frost-tender.
Billbergia		
nutans	Queen's tears	Evergreen, 0.2 m, well-drained peaty soil, sun.
Bougainvillea	various	Evergreen climber, most rich drained soils, sun, frost-tender.
Brachyglottis		
repanda		Evergreen, 6 m, most soils, sun, from New Zealand.
Brahea		
brandegeei	San Jose palm	Evergreen, 9 m, rich sandy loam, sun.
dulcis	Mexican palm	Evergreen, 6 m, rich sandy loam, sun.
Bupleurum		
fruiticosum		Evergreen, 2 m, well-drained alkaline soil, sun.
Buxus		
sempervirens	Box	Evergreen, 8 m, well-drained alkaline, sun. Hedge, medicinal.
Caesalpina		
regia	Flamboyant tree	Deciduous, 12 m, well-drained, sun.
Calamintha		
grandiflora	Calamint	Perennial, 0.5 m, well-drained, soil, sun.
nepeta	Calamint	Perennial, 0.6 m, well-drained, soil, sun.
officinalis	Mountain thyme	Perennial, 0.5 m, well-drained, soil, sun.
Calandrina		
umbellata	Rock purslane	Perennial, light to medium soil, well-drained, sun.
Callicarpa		
dichotoma	Beauty berry	Deciduous, 2 m, most soils, sun or shade.
Carissa		
ovata	Carissa	Evergreen, 7 m, most soils, sun or shade.
Carpinus		
betulus	Witch hazel	Deciduous, 25 m, light soil, well-drained, sun.
caroliniana	American hornbeam	Deciduous, 8 m, well-drained loam, sun.
japonica	Japanese hornbeam	Deciduous, 10 m, well-drained, sun.
Carpobrotus	Pigface	Various. Perennial succulent, 0.3 m, some frost-tender.
Ceanothus		
americanus	Jersey tea	Deciduous, 1.2 m, well-drained, sun.
cyaneus	Californian lilac	Evergreen, 2 m, well-drained, sun.
dentatus	Dwarf blue ceanothus	Evergreen, 3 m, most soils, sun.
rigidus	Monterey ceanothus	Evergreen, 3 m, light to medium soil, sun.
roweanus		Evergreen, 1 m, most soils, sun.
Cedrela		
sinensis	Chinese cedar	Deciduous, 12 m, most soils, sun or shade.

Cedrus		
deodara	Deodar cedar	Evergreen, 30 m, most soils, sun or shade.
libani	Cedar of Lebanon	Evergreen, 30 m, most soils, sun.
Celastrus		
hypoleucus	Waxworks	Deciduous vine, most soils, sun or shade.
orbiculatus	False bittersweet	Deciduous vine, most soils, sun or shade.
scandens	American bittersweet	Deciduous vine, most soils, sun or shade.
Celtis		
australis	Nettle tree	Deciduous, 25 m, alkaline soil, sun.
Centaurea		
cyanis	Cornflower	Annual, 0.9 m, well-drained, sun
scabiosa	Greater knapweed	Perennial, 1 m, well-drained, sun.
Cereus		
grandiflorus	Night-blooming cereus	Cactus, 5 m, sand, sun, tropical.
jamacaru		Cactus, 9 m, sand, sun, tropical.
peruvianus		Cactus, 3 m, sand, sun, tropical.
Choisya		
ternata	Mexican orange-blossom	Evergreen, 2 m, well-drained loam, sun.
Chrysalidocarpus		
lutescens	Cane palm	Evergreen, 10 m, most soils, sun, frost-tender.
Chrysanthemum		
leucanthemum	Ox-eye daisy	Perennial, 0.7 m, most soils, sun.
parthenium	Feverfew	Perennial, 0.6 m, most soils, sun.
vulgare		Perennial, 1 m, most soils, sun.
Cistus		
albidus	Rock rose	Evergreen, 2 m, sandy, sun.
crispus	Rock rose	Evergreen, 1 m, sandy, sun, frost-tender.
hirsutus	Sun rose	Evergreen, 1 m, sandy, sun.
incanus	Mediterranean rose	Evergreen, 1 m, sandy, sun.
ladaniferus	Sun rose	Evergreen, 1.5 m, sandy, sun.
lauriflorus		Evergreen, 2.5 m, sandy, sun.
salviifolius		Evergreen, 1 m, sandy, sun.
villosus		Evergreen, 1 m, sandy, sun.
Citrus		
auranteum	Seville orange	Evergreen, 15 m, most soils, sun.
bergamia	Bergamot	Evergreen, 3 m, well-drained, sun.
reticulata	Tangerine	Evergreen, 4 m, well-drained, sun.
trifoliata	Japanese bitter orange	Evergreen, 3 m, most soils, sun.
Colutea		
arborescens	Bladder senna	Deciduous, 5 m, most soils, sun.
istria	Bladder senna	Deciduous, 2 m, most soils, sun.
Comptonia		
peregrina	Meadow fern	Deciduous, 1.5 m, light to medium soil, well-drained, sun.

Coreopsis		
gigantea	Tickseed	Annual, 2 m, most soils, sun.
grandiflora	Tickseed	Perennial, 0.6 m, most soils, sun.
rosea	Calliopsis	Perennial, 0.6 m, most soils, sun.
Cotoneaster		
acutifolius	Rock spray	Deciduous, 3 m, most soils, sun or shade.
adpressus		Deciduous, 0.5 m, most soils, sun or shade.
amoenus		Deciduous, 1.5 m, most soils, sun or shade.
bullatus		Deciduous, 4 m, most soils, sun or shade.
congestus		Deciduous, 0.6 m, most soils, sun or shade.
conspicuus		Deciduous, 2 m, most soils, sun or shade.
dammeri		Evergreen, 0.3 m, most soils, sun or shade.
dielsianus		Semi-deciduous, 2.5 m, most soils, sun.
franchetti		Evergreen, 2.5, light to medium, sun.
frigidus		Deciduous, 6 m, most soils, sun or shade.
harrovianus		Evergreen, 2 m, most soils, sun.
henryanus		Evergreen, 3 m, most soils, sun or shade.
horizontalis		Deciduous, 0.6 m, gravel soils, sun.
intergerrimus		Deciduous, 2 m, gravel soils, sun.
lacteus		Evergreen, 4 m, most soils, sun or shade.
lanatus		Evergreen, 2 m, well-drained, sun.
melanocarpus		Deciduous, 2 m, most soils, sun or shade.
microphyllus		Evergreen, 1.5 m, well-drained, sun.
pannosus		Evergreen, 3 m, most soils, sun or shade.
prostratus		Evergreen, 3 m, most soils, sun or shade.
racemiflorus		Semi-evergreen, 2.5 m, most soils, sun or shade.
rotundifolius		Semi-evergreen, 2.5 m, well-drained, sun.
rugosus		Evergreen, 2.5 m, most soils, sun.
salicifolius		Evergreen, 4 m, most soils, sun or shade.
serotinus		Evergreen, 5 m, most soils, sun or shade.
simonsii		Semi-evergreen, 2.5 m, most soils, sun or shade.
Crassula		
arborescens	Silver jade plant	Evergreen, 4 m, rich sand, partial shade.
argentea		Perennial, 1 m, rich sand, partial shade.
barbata		Evergreen, 10 cm, rich sand, sun.
brevifolia		Evergreen, 0.2 m, rich sand, partial shade.
Crataegus		
monogyna	English hawthorn	Deciduous, 8 m, most soils, sun, must be kept under control.
oxyacantha	Hawthorn (white flower)	Deciduous, 5 m, most soils, sun, must be kept under control.

Cupressus		
abramsiana	Santa Cruz cypress	Evergreen, 20 m, most alkaline soils, sun.
arazonica	Smooth Arizona cypress	Evergreen, 20 m, most alkaline soils, sun.
bakerii	Modoc cypress	Evergreen, 15 m, most alkaline soils, sun.
cashmeriana	Cashmere cypress	Evergreen, 20 m, most soils, sun.
funebris	Chinese weeping cypress	Evergreen, 20 m, most soils, sun.
glabra	Arizona cypress	Evergreen, 15 m, most soils, sun.
goveniana	Californian cypress	Evergreen, 15 m, most soils, sun.
lusitanica	Mexican cypress	Evergreen, 30 m, most soils, sun.
sempervirens	Italian cypress	Evergreen, 30 m, light poor soil, sun.
torulosa	Bhutan cypress	Evergreen, 20 m, most soils, sun.
Cycas		
revoluta	Dwarf sago palm	Evergreen fern, 3.5 m, poor soils, sun, wind-tender.
Cytisus		
albus	White Spanish broom	Evergreen, 3 m, light to medium soil, sun.
battandieri	Moroccan broom	Deciduous, 3 m, light acid drained sand, sun.
canariensis	Canary broom	Evergreen, 2 m, poor dry soils, sun.
linifolius	Flax-leaf broom	Evergreen, 1.5 m, light to medium soil, well-drained, sun, frost-tender.
madeiriensis	Madeira broom	Evergreen, 6 m, light to medium soil, well-drained, sun.
monspessulanus	Montpellier broom	Evergreen, 3 m, acid, well-drained, sun.
multiflorus	Portugal broom	Evergreen, 4 m, poor dry soils, sun.
procumbens	Balkan broom	Evergreen, 1 m, acid soil, drained, sun.
pungens		Evergreen, 1 m, light to medium soil, sun.
racemosus	Fragrant broom	Evergreen, 2 m, light to medium soil, sun.
radiatus	Royal broom	Evergreen, 0.6 m, most acid soils, sun.
scoparius	Scotch broom	Evergreen, 3 m, light to medium acid soil, sun.
Delonix		
regia	Royal poinciana	Deciduous, 12 m, most soils, sun.
Delosperma	Various	Perennial, less than 0.5 m, well-drained, sun. Succulent.
Deutzia		
gracilis	Japanese snowflower	Deciduous, 1.5 m, most soils, sun or shade.
Dianthus		
alpinus	Pink	Perennial, 0.2 m, light alkaline soil, well-drained, sun.
barbatus	Sweet William	Perennial, 0.6 m, light to medium soil, sun.
caryophyllus	Cheddar pink	Perennial, 0.3 m, light to medium soil, sun.
chinensis		Annual, 0.75 m, light to medium soil, sun.
deltoides	Maiden pink	Perennial, 0.6 m, light to medium soil, sun.
plumaris	Cottage pink	Perennial, 0.4 m, light to medium soil, sun.
Diospyros		
kaki	Japanese persimmon	Deciduous, 12 m, most soils, sun or shade.

Dorianthes		
palmerii	Palmer spear lily	Evergreen, 5 m, well-drained, succulent.
Dudleya	Various	Succulents, up to 1.5 m, well-drained, sun, spiky.
Echeveria	Various	Succulents, less than 0.5 m, well-drained, sun.
Elaeagnus		
augustifolia	Russian olive	Deciduous, 5 m, dry soil, well-drained, sun or shade.
commutata	Silverberry	Deciduous, 3 m, light to medium soil, sun, windprone.
Ephreda		
vulgaris	Teamsters tea	Evergreen, 2 m, sandy soil, sun, medicinal.
Erica		
baccans	Berry heath	Evergreen, 1.5 m, medium to heavy soil, sun, frost-tender.
bauera	Bridal heath	Evergreen, 1 m, acid soils, well-drained, sun, frost-tender.
canaliculata	Purple heath	Evergreen, 2 m, light to medium soil, well-drained, sun.
carnea	Mountain heath	Evergreen, 0.3 m, well-drained, sun.
cruenta	Blood-red heath	Evergreen, light to medium, well-drained, sun.
curviflora	Amber heath	Evergreen, 1.5 m, light to medium, acid soil, sun.
glandulosa	Spotted heath	Evergreen, 1.5 m, light to medium soil, sun.
gracilis	Slender heath	Evergreen, 0.6 m, light to medium acid soil, sun.
lusitanica	Portugese heath	Evergreen, 3 m, well-drained, sun, frost-tender.
oatesii	Winter gem heath	Evergreen, 0.6 m, light to medium soil, well-drained, sun.
persoluta	Pink garland heath	Evergreen, 1 m, well-drained, sun.
perspicua	Prince of Wales heath	Evergreen, 1.5 m, light to medium soil, well-drained, sun.
sessiliflora	Green trumpet heath	Evergreen, 1 m, light to medium soil, well-drained, sun.
speciosa	Showy heath	Evergreen, 1 m, light to medium soil, well-drained, sun.
ventricosa	Wax heath	Evergreen, 1.5 m, light to medium soil, well-drained, sun, frost-tender.
versicolor	Rainbow heath	Evergreen, 1.5 m, light to medium soil, well-drained, sun.
verticillata	Scarlet heath	Evergreen, 1.5 m, light to medium soil, well-drained, sun, frost-tender.
vestita	Trembling heath	Evergreen, 1 m, light to medium soil, well-drained, sun.
Erigeron		
auranticus	Golden fleabane	Evergreen, 0.45 m, most soils, sun or shade.
canadensis	Canadian fleabane	Annual, 1 m, most soils, sun or shade, medicinal.

Eriobotrya		
japonica	Loquat	Evergreen, 5 m, rich heavy soils, sun, needs some water.
Erodium	Various	
Erysimum		
asperum	Alpine wallflower	Perennial, 1 m, most soils, sun or shade.
Erythrina		
americana	Coral tree	Deciduous, 10 m, light to medium soil, well-drained, sun, frost-tender.
corallodendrum	Coral tree	Deciduous, 4 m, most soils, sun.
Eunymus		
fortunei		Evergreen climber, 6 m, most soils, sun or shade.
japonica	Evergreen spindle tree	Evergreen, 2.5 m, most soils, sun or shade.
Euphorbia	Various	Some frost-tender.
Euryops		
abrotanifolius		Evergreen, 2 m, light sandy soil, well-drained, sun, frost-tender.
athanasiae	Marguerite Othanna	Evergreen, 1 m, light sandy soil, well-drained, sun.
tenuissimus	Paris Daisy	Evergreen, 1.5 m, light sandy soil, well-drained, sun, frost-tender.
Ferocactus	Various	Cactus, up to 3 m, well-drained, sun, spiky.
Foeniculum		
vulgare	Fennel	Perennial, 1.5 m, most soils, sun.
Ficus		
palmeri		Evergreen, 4 m, most soils, sun or shade.
sycamorus	Egyptian sycamore	Deciduous, 13 m, most soils, sun.
Foeniculum		
vulgare	Fennel	Perennial, 1.8 m, most soils, sun, bronze version available.
Fouquieria	Various	Succulent, up to 5 m, well-drained, sun.
Fraxinus		
lanceolata	Green ash	Deciduous, 13 m, most soils, sun or shade.
ornus	Manna ash	Deciduous, 10 m, deep soil, well-drained, sun.
syriaca	Syrian ash	Deciduous, 8 m, most soils, sun.
velutina	Velvet ash	Deciduous, 10 m, alkaline soils, sun.
Fremontia		
californica	Flannel bush	Semi-deciduous, 5 m, light to medium soil, well-drained, sun.
mexicana	Flannel bush	Evergreen, 4 m, light to medium soil, well-drained, sun.
Gardenia		
thunbergia	Thunberg gardenia	Evergreen, 2.5 m, light to medium soil, well-drained, sun, frost-tender.
Gazania	Various	Perennial, 0.3 m, most soils, sun.

Gelsemium		
nitidum	Carolina jasmine	Evergreen vine, rich soil, well-drained, sun.
sempervirens	Jessamine	Evergreen vine, rich soil, well-drained, sun.
Genista		Not recommended because they become weeds.
Geranium	Various	Perennial, 1 m, most drained soils, sun, frost-tender.
Gleditsia		
triacanthos	Honey locust	Deciduous, 30 m, most soils, sun, thornless variety available.
Halimium		
atricifolium	Sun rose	Evergreen, 1.5 m, most soils, sun.
halimifolium		Evergreen, 1 m, most soils, sun.
lasianthum	Sun rose	Evergreen, 1 m, most soils, sun.
Hedeoma		
pulegioides	Pennyroyal	Annual, 0.15 m, most soils, sun or shade.
Helianthemum		
canadensis	Rock rose	Perennial, 0.3 m, alkaline sand, sun.
chamaecistus	Rock rose	Evergreen, 0.6 m, alkaline sand, sun.
Helianthus		
annua	Sunflower	Annual, 1.75 m, rich soil, sun.
tuberosus	Jerusalem artichoke	Annual, 1.5 m, rich soil, sun.
Helichrysum		
angustifolium	Curry plant	Perennial, 0.3 m, most soils, sun.
petiolatum		Perennial, 1.2 m, light to medium soil, sun.
Helleborus		
foetidus	Hellebore	Evergreen, 0.45 m, most drained soils, sun.
Hesperis		
matronalis	Rocket	Biennial, 0.9 m, most drained soils, sun, culinary.
Hibiscus		
tiliaceus	Cottonwood	Evergreen, 9 m, light soils, sun, frost-tender.
Hymenosporum		
flavum	Native frangipani	Evergreen, 8 m, light to medium soil, sun, frost-tender when young.
Hyoscyamus		
aureus	Yellow henbane	Annual, 0.7 m, most soils, sun.
Hypericum		
androsaemum	Amber	Evergreen, 1 rn, light to medium soil, well-drained, shade.
Hyssopus		
officinalis	Hyssop	Evergreen, 0.6 m, light soil, well-drained, shade.
iberis		
amara	Bitter candytuft	Annual, 0.3 m, most soils, sun or shade.
gilbraltarica	Gibraltar candytuft	Perennial, 0.3 m, most soils, sun or shade.

umbellata	Globe candytuft	Annual, 0.4 m, most soils, sun or shade.
Jacaranda		
acutifolia		Deciduous, 15 m, rich soil, well-drained, sun, frost-tender, hard to find.
Jasminum		
rex	King jasmine	Evergreen, 2.5 m, rich soil, well-drained, semi-shade, semi-tropical.
Juniperus		
communis	Juniper	Evergreen, 2 m, acid sand, sun.
deppeana	Alligator juniper	Evergreen, 15 m, most soils, sun or shade.
drupacea	Syrian juniper	Evergreen, 20 m, well-drained, sun.
occidentale	Western juniper	Evergreen, 20 m, well-drained, sun.
osteosperma	Utah juniper	Evergreen, 7 m, well-drained, sun.
oxycedrus	Prickly juniper	Evergreen, 8 m, light to medium acid soil, sun.
sabina	Savin juniper	Evergreen, 3 m, alkaline soil, well-drained, sun.
Koelreuteria		
elegans	Flame gold	Deciduous, 20 m, rich soil, well-drained, sun.
paniculata	Golden rain tree	Deciduous, 6 m, rich soil, well-drained, sun.
Laburnum		
anagyroides	Golden chain	Deciduous, 6 m, alkaline soil, well-drained, sun.
Lamprathus	Pigface	Various succulents, less than 0.5 m, well-drained, sun.
Lantana		Don't use in NSW, NT and Qld where it is invasive (although hybrids are safer).
camara		Evergreen, 3 m, most soils, sun or shade. (Noxious weed in Qld and NSW.)
montevidensis		Evergreen, 0.5 m, most soils, sun or shade.
rugulosa		Evergreen, 2 m, most soils, sun or shade.
salviifolia		Evergreen, 2 m, most soils, sun or shade.
sanguinea		Evergreen, 1.5 m, most soils, sun or shade.
tiliifolia		Evergreen, 2 m, most soils, sun or shade.
Larix		
americana	Larch	Deciduous, 30 m, most soils, sun or shade.
Lathyrus		
grandiflorum	Wild pea	Evergreen vine, most drained soils, sun.
laetiflorus	Pink perennial pea	Evergreen vine, most drained soils, sun.
pratensis	Meadow vetchling	Evergreen vine, most drained soils, sun.
pubescens	Blue perennial pea	Evergreen vine, most drained soils, sun.
splendens	Pride of California	Evergreen vine, most drained soils, sun.
sylvestris	Everlasting pea	Evergreen vine, most drained soils, sun.
Laurus		
canariensis	Canary Island laurel	Evergreen, 20 m, rich soil, well-drained, sun.
Lavandula		
angustifolia	English lavender	Evergreen, 1.5 m, light to medium alkaline soil, sun.

dentata	French lavender	Evergreen, 1.5 m, light to medium alkaline soil, sun.
officinalis	Lavender	Evergreen, 0.6 m, light to medium alkaline soil, sun.
spicata	True lavender	Evergreen, 1 m, light to medium alkaline soil, sun.
stoechas	Topped lavender	Evergreen, 0.6 m, light to medium alkaline soil, sun. (Noxious weed in Vic.)
Liatris		
aspera	Gay feather	Perennial, 2 m, most soils, sun or shade.
odoratissima	Deers tongue	Perennial, 1.5 m, most soils, sun or shade.
scariosa	Blazing star	Perennial, 1 m, most soils, sun or shade.
Ligustrum		
vulgare	Privet	Evergreen, 5 m, most soils, sun or shade.
Linaria		
cymbalaria	Ivy-leafed Toadflax	Perennial, 0.3 m, most soils, sun.
vulgaris	Toadflax	Perennial, 1 m, most soils, sun.
Linum	Various flaxes	Snail problems and fire risk, most soils.
Lonicera		
caprifolium	Sweet woodbine	Deciduous vine, most soils, sun.
flava	Yellow honeysuckle	Deciduous vine, most soils, sun.
fragantissima	Winter honeysuckle	Semi-evergreen, 2.5 m, most soils, sun.
periclymenum	Honeysuckle	Evergreen vine, most soils, sun.
splendida	Pink woodbine	Evergreen vine, most soils, sun.
Lupinus		
album	Lupin	Annual, 1 m, most soils, sun.
polyphyllus	Lupin	Perennial, 1.5 m, most soils, sun.
Maclura		
pomifera	Osage orange	Deciduous, 20 m, most soils, sun or shade.
Melianthus		
major	Large honey flower	Evergreen, 3 m, rich soil, sun.
Mimulus		
aurantiacus	Monkey musk	Perennial, 0.3 m, most soils, sun.
glutinosus	Monkey musk	Evergreen, 1.2 m, most soils, sun.
moschatus	Monkey musk	Perennial, 0.3 m, most soils, sun.
Morus		
nigra	Black mulberry	Deciduous, 15 m, most soils, sun.
rubra	Red mulberry	Deciduous, 25 m, most soils, sun.
Nandina		
domestica	Sacred bamboo	Evergreen, 2 m, most soils, sun.
Nepeta		
mussini	Catmint	Perennial, 0.3 m, most soils, sun.
Nerium		
oleander	Oleander	Evergreen, 4 m, most soils, sun, leaves and stems poisonous.

Oenothera		
biennis	Evening primrose	Biennial, 1.5 m, most soils, sun or shade.
Olea		
africana	African olive	Evergreen, 8 m, rich soil, well-drained, sun, frost-tender.
europea	European olive	Evergreen, 7 m, rich soil, well-drained, sun, frost-tender.
glandulifera		Evergreen, 20 m, rich soil, well-drained, sun, frost-tender.
laurifolia	Black ironwood	Evergreen, 25 m, light soil, well-drained.
Opuntia		
vulgaris	Prickly pear	Not recommended.
Osteospermum		
amplectans	Yellow velvet daisy	Perennial, 1 m, well-drained, sun, frost-tender.
barberiae	Pink velvet daisy	Perennial, 1 m, well-drained, sun, frost-tender.
ecklonis	White velvet daisy	Evergreen, 1 m, well-drained, sun, frost-tender.
fruticosum	Pink velvet daisy	Evergreen, 1 m, well-drained, sun, frost-tender.
Pelargonium	Various	Can be frost-tender.
Penstemon	Various	Said to be drought-tender, but has adapted well to southern Australia.
Phlomis		
fruticosa	Jerusalem sage	Evergreen, 1.2 m, light to medium soil, well-drained, sun, salt-tolerant.
Phoenix		
canariensis	Canary Island date palm	Evergreen, 15 m, most soils, sun.
dactylifera	Date palm	Evergreen, 30 m, most soils, sun, frost-tender.
reclinata	Senegal date palm	Evergreen, 8 m, most soils, sun.
Phormium	Flax	Hardy but can be a problem.
Pinus		
attenuata	Knob cone pine	Evergreen, 25 m, acid soil, well-drained, sun.
coulteri	Big cone pine	Evergreen, 30 m, acid soil, well-drained, sun.
halepensis	Aleppo pine	Evergreen, 12 m, most soils, sun.
muricata	Bishop pine	Evergreen, 30 m, acid soil, well-drained, sun.
peuce	Macedonian pine	Evergreen, 30 m, most drained soil, sun.
ponderosa	Ponderosa	Evergreen, 70 m, most drained soil, sun.
radiata	Monterey pine	Evergreen, 20 m, most drained soil, sun.
torreyana	Soledad pine	Evergreen, 12 m, most drained soil, sun.
Plumbago		
capensis	Cape plumbago	Evergreen, 2 m, most drained soil, sun.
europea	Plumbago	Evergreen vine, most drained soil, sun or shade.
Poncirus		
trifoliata	Japanese bitter orange	Evergreen, 5 m, most drained soil, sun or shade.
Polygonum	Various	Most soils.
Populus		
trichocarpa	Black cottonwood	Deciduous, 70 m, most drained soil, sun.

Portulaca	Various	
Potentilla		
tabernaemontana	Spring cinquefoil	Perennial, 0.2 m, light to medium soil, drained, sun.
Prosopis		
glandulosa	Honey mesquite	Evergreen, 5 m, light to heavy soil, sun.
Protea	Various	Good value, need good drainage and sun.
Prunus		
hortulana	Goose plum	Deciduous, 10 m, light to medium soil, sun.
lusitanica	Portugal laurel	Evergreen, 6 m, light to medium soil, sun.
Quercus		
agrifolia	Coastal live oak	Evergreen, 35 m, deep rich soils, sun.
canariensis	Algerian oak	Evergreen, 25 m, medium to heavy soils, sun.
castanaefolia	Chestnut-leafed oak	Deciduous, 25 m, most soils, sun or shade.
cerris	Turkey oak	Deciduous, 35 m, most soils, sun or shade.
chrysolepis	Goldcup oak	Evergreen, 20 m, most soils, sun or shade.
douglasii	Blue oak	Deciduous, 20 m, most soils, sun or shade.
engelmanii		Evergreen, 20 m, most soils, sun or shade.
garryana	Garry oak	Deciduous, 30 m, most soils, sun.
kelloggii	Californian black oak	Deciduous, 28 m, most soils, sun or shade.
libani	Lebanon oak	Semi-evergreen, 12 m, most soils, sun.
lusitanica	Portugal oak	Deciduous, 15 m, most soils, sun.
maralandica	Blackjack oak	Deciduous, 10 m, poor soil, well-drained, sun.
pontica	Armenian oak	Deciduous, 10 m, most soils, sun or shade.
robur	English oak	Deciduous, 30 m, light alkaline soil, well-drained, sun.
stellata	Post oak	Deciduous, 20 m, most soils, sun.
suber	Cork oak	Deciduous, 20 m, sandy acid soil, sun.
Raoulia		
haastia		Perennial, 0.5 m, rocky soil, sun.
mammillaris	Vegetable sheep	Perennial, 0.2 m, rocky soil, sun.
Ricinus		
communis	Castor oil plant	Evergreen, 3 m, most soils, sun.
Robinia		
pseudoacacia	Black locust	Deciduous, 25 m, most soils, sun. Suckers and has various grafted hybrids.
Rosa		
canina	Dog rose	Evergreen, 5 m, loamy soil, sun.
setigera	Prairie rose	Deciduous, 2 m, loamy soil, sun.
Rosmarinus		
officinalis	Rosemary	Evergreen, 1.5 m, well-drained, sun.
Rubus		
idaeus	Raspberry	Perennial, 1.5 m, well-drained, sun.
parvifolus	Native bramble	Evergreen, 0.6 m, well-drained, sun.
strigosus	Raspberry	Biennial, 2.5 m, well-drained, sun.

Ruscus		
aculeatus	Box Holly	Evergreen, 1 m, most soils, sun or shade.
Salvia		
aurea	Golden sage	Evergreen, 1.2 m, sandy well-drained soil, sun, frost-tender.
grahamii		Evergreen, 2 m, composted soil, well-drained, sun, frost-tender.
involucrata	Mexican sage	Evergreen, 1.2 m, most soils, sun, frost-tender.
lavandulifolia	Spanish sage	Evergreen, 0.5 m, most soils, sun, frost-tender.
leucantha	Mexican mintbush	Evergreen, 0.6 m, light to medium soil, sun.
officinalis	Garden sage	Evergreen, 0.7 m, alkaline soils, sun.
sclarea	Clary sage	Evergreen, 1 m, alkaline soils, sun.
Santolina		
chamaecyparissus	Cotton lavender	Deciduous, 1 m, well-drained, sun.
virens		Deciduous, 1 m, well-drained, sun.
Schinus		
molle	Peppercorn	Evergreen, 15 m, well-drained, sun.
Sedum	Various perennials	Well-drained.
Selenicereus		
grandiflorus	Queen of night	Cactus, 5 m, most soils, sun.
Senicio		
ericulifolius	Hoary groundsel	Perennial, 1 m, light to medium soil, drained, sun.
maritima	Dusty miller	Evergreen, 10 m, light to medium soil, sun, frost-tender.
pendula		Perennial, 0.5 m, light to medium soil, sun.
Silene		
nutans	Nottingham catchfly	Perennial, 0.5 m, sandy soil, sun.
Sparteum		
junceum	Spanish broom	Deciduous, 3 m, most soils, sun.
Tamarix		Do not plant in northern Australia, where it is invasive.
aphylla	Athel tree	Evergreen, 6 m, most soils, sun, accumulates salt in leaves.
Taxus		
baccata	Yew	Evergreen, 20 m, moist alkaline soils, sun.
cuspidata	Japanese yew	Evergreen, 20 m, alkaline soils, sun.
Teucrium		
betonicum	Germander	Evergreen, 0.6 m, well-drained, sun, frost-tender.
chamaedrys	Wall germander	Perennial, 0.45 m, well-drained, sun or shade.
fruticans	Bush germander	Evergreen, 1.2 m, well-drained, sun, frost-tender.
marum	Cat thyme	Perennial, 1.2 m, well-drained, sun.
scorodonia	Germander	Perennial, 0.6 m, well-drained, shade.
Thymus		Various thyme are happy in well-drained soils and sun.

Ulex		
europaefus	Gorse	Noxious weed
Valeriana		Said to be drought-tender but has adapted well in southern Australia.
officinalis	Valarian	Perennial, 1.5 m, most soils, sun.
Verbascum	Mullein, various	Biennial, 2 m, most soils, sun.
Viburnum		
tinus	Laurestinus	Evergreen, 3 m, most soils, sun.
Vinca		
major	Periwinkle	Evergreen vine, 3 m, most soils, sun or shade.
minor	Periwinkle	Evergreen creeper, most soils, sun or shade.
Virgilia		
oroboides	Cape Virgilia	Evergreen, light soil, well-drained, sun.
Washingtonia		
filifera	Fan palm	Evergreen palm, 12 m, well-drained, sun.
robusta	Mexican fan palm	Evergreen palm, 25 m, well-drained, sun.
Yucca		
aloifolia	Dwarf yucca	Evergreen, 15 m, well-drained, sun.
brevifolia	Joshua tree	Evergreen, 4 m, well-drained, sun.
filamentosa	Adams needle	Evergreen, 1.5 m, well-drained, sun or shade.
glauca	Dwarf yucca	Evergreen, 0.6 m, well-drained, sun or shade.
gloriosa	Mound lily	Evergreen, 1 m, well-drained, sun or shade.
recurvifolia	Weeping yucca	Evergreen, 2 m, well-drained, sun or shade.
whipplei	Our Lords needle	Evergreen, 2 m, well-drained, sun or shade.
Zizyphus		
vulgaris	JuJube	Evergreen, 8 m, well-drained, rich soil, sun.

Appendix 4

Fire-retardant species

Green lawns do not burn, but neither do they provide a shield from the intense heat of fire. Deciduous shrubs and trees form the best vegetative shields as they mostly do not burn readily. Instead, they tend to shrivel as the water in the leaves and branches boils and evaporates. Only then is there a chance of them burning, and by that time most fires have passed or are under control. They will not look good, but you will have saved your buildings.

The best fire defence plan is to have lawn or a paved area close to buildings with the fire-shield plants further out, and the more flammable vegetation further out again. This approach fits in neatly with the concept of keeping the wet areas closer to the house to minimise plumbing costs.

The plants listed in this table are mostly Australian natives preferring good drainage, and are drought- and frost-hardy unless otherwise stated. Many will need dry wood pruned off as they age.

Ground covers

Aptenia sp.	Noon-flowers native of South Africa, frost-tender, bright daisy-like flowers.
Carpobrotus sp.	Pigface natives of various states, Cacinaciformis native of South Africa, bright flowers, edible fruit.
Dichondra repens	Kidney weed is an excellent grass substitute for low traffic areas, requires some water, never needs mowing.
Kennedia prostrata	Running Postman, delightful red spring flowers, for cool temperate areas.
Lampranthus sp.	Ice plant native of South Africa, sunny site, bright daisy-like flowers.
Myoporum parvifolium	Creeping boobiallia, likes sunny open site, white flowers, old growth can burn unless thinned.

Shrubs and trees

Acacia baileyana	Cootamundra wattle, to 10 m, smoky feathery foliage.
Acacia dealbata	Silver wattle, to 10 m, quick grower, silver foliage.
Acacia howittii	Sticky wattle, to 8 m, tolerates poor drainage, good windbreak, keep 2 m from drains.
Acacia melanoxylon	Blackwood, to 30 m, erect, slow grower, for wet temperate climates.
Acacia vestita	Hairy wattle, to 4 m, semi-shade, weeping grey foliage.
Allocasuarina verticillata	Drooping sheoak, delightful weeping needle-like foliage, leaf fall is flammable.
Angophora costata	Smooth-barked apple, beautiful pink-hued smooth bark, for large gardens.
Artemisia sp.	Wormwoods and southernwoods, up to 1.5 m, make soft hedges, generally soft grey foliage.
Atriplex nummularia	Old man saltbush, to 3 m, makes a good dry-tolerant hedge.
Atriplex rhagodioides	Silver saltbush, to 1 m, makes a good hedge, has silvery-grey leaves.
Atriplex vesicaria	Bladder saltbush, to 1 m, silvery-grey leaves.
Banksia marginata	Silver banksia, to 4 m, leaves blunt-ended with silver undersides, creamy flowers.
Banksia ornata	Desert banksia, to 3 m, grey leaves, flowers dull yellow.
Casuarina cunninghamiana	River sheoak Self-mulching drooping foliage, flammable leaf fall, inclined to sucker, for large gardens.
Ceratonia siliqua	Carob, to 10 m, native of Syria, likes alkaline soils, has pods which can be used to make mock chocolate or stock feed.
Chamaecytisus prolifer	Tree lucerne, to 5 m, native of Canary Islands, does not like extreme frosts, grey-green foliage, white spring flowers followed by pods, good hedge plant, not all that attractive, but seed drop and hedge clippings good fodder for poultry run or other livestock, best for country property or city farm.
Corymbia maculata	Spotted gum, lovely smooth bark changes colour with the seasons, for large gardens, frost-tender when young.
Cotinus obovatus	American smoke bush, to 6 m, native of North America, prefers shelter and wet climate, pale lavender flowers.
Crataegus laevigata	Pauls Scarlet, Scarlet hawthorn, grows to 10 m, native of Europe, heavy soils, pink and white flowers, best as a hedge, needs control as a weed.
Dodonea viscosa	Hop bush, to 2 m, mainly for indigenous gardens.
Einadia sp.	Salt bushes, low shrubs
Eucalyptus bauerana*	Blue box, to 25 m, likes moist climates, bluish-grey leaves.
Eucalyptus gummifera*	Blood wood, to 40 m, long tapering leaves.
Eucalyptus maculata*	Spotted gum, to 50 m, various soils, frost-tender when young, beautiful bark.
Grevillea rosmarinifolia	Rosemary grevillea, to 2 m (also dwarf to 1 m), good hedge, good bird habitat.
Laurus nobilis	Sweet bay, to 12 m, most soils, some irrigation, leaves used in cooking and as silverfish repellent.
Myoparium insulare	Boobialla, to 5 m, wind-resistant, old growth needs trimming, large hedge.
Myoparium montanum	Waterbush, to 3 m.
Photinia sp.	Photinias, natives of China and Asia, likes wetter climates, can cause allergies.
Senna artemisioides	Desert cassia, shrub to 2 m.

*Eucalypts listed here are the least flammable types, but the foliage still burns. Suit large gardens.

Appendix 5

Drain stranglers and cloggers

Drain breakers, stranglers or cloggers should be kept well away from all underground pipes, including your neighbours' pipes - more than 3 m in clay soil and 4 m in sandy soil. The following list is based on advice from authorities but it is not complete, so you should ask your nursery or local council for advice. The list includes large-rooted species that displace pipes and even break them, as well as fibrous rooted plants that penetrate any joints.

Acacia (wattle)	acuminata, cultriformis, cyclops, howittii, iteaphylla, longifolia, microbotrya, pendula, retinodes, salicina, saligna, sowdenii, terminalis, trineurad, vernicjlua, victoriae.
Acer (maple)	negunda, platanoides, pseudoplatanus, sacharum.
Acmena	smithii (lilly pilly).
Actinostrobus	pyramidalis (Swan River cypress).
Aesculus (chestnut)	hippocastanum, indica.
Agonis	flexuosa (peppermint tree or willow myrtle).
Brachychiton	acerfolius (Illawarra flame tree), discolor (white kurrajong), populneus (kurrajong).
Callistemon (bottlebrush)	acuminatus, citrinus, formosus, Harkness, lilacinus, macropunctatus, phoeniceus, rigidus, salignus, viminalis (Captain Cook), violaceus.
Callitris (native pine)	columellaris, huegelii, preissii.
Calothamnus	asper (rough-leaf netbush).
Castanea	sativa (Spanish chestnut).
Casuarina (native oak)	cristata, cunninghamiana, glauca, stricta, torulosa.
Cedrus	deodora (deador cedar).
Ceratonia	siliqua (carob).
Cercis	siliquastrum (Judas tree).
Coprosma	quadrifida (prickly currant), repens (mirror bush).

Eucalyptus	most over 3 m high when mature.
Fagus	silvatica (European beech).
Ficus	elastica (Indian rubber fig), benghalensis (banyan), benjamina (weeping fig), indica (Indian).
Fraxinus (ash)	alba, angustafolia, excelsior, excelsior 'Aurea', oxycarga, Raywoodi.
Grevillea	robusta.
Hakea	elliptica, eriantha, kippistiana, laurina, leucoptera, oleofolia, petiolaris, salicifolia, saligna, suaveolens, sulcata, undulata, victoriae.
Homalanthus	populifolius (bleeding heart tree).
Jacaranda	mimosifolia (jacaranda).
Leptospermum (tea tree)	brachyandrum, ericoides, grandiflorum, juniperinum, laevigatum, longifolium, petersonii, squarrosum.
Liquidambar	formosana, orientalis, styraciflua.
Malus (Crabapple and domestic apple fruit tree)	
Melaleuca (paperbark)	acuminata, argentia, armillaris, decora, fulgens, glomerata, halmaturorum, huegelii, incana, lanceolata, lateritia, linarifolia, nesophila, pauciflora, pentagona, radula, sieberi, squamea, styphelioides, symphocarpa, viridifolia.
Melia azedarach	
Myoporum	insulare, montannum.
Pittosporum	phillyreoides, rhombifolium, undulatum (sweet pittosporum).
Populus	fremonth (cottonwood), nigra (Lombardy poplar).
Prunus	x blireana (double-flowering plum), domestica(plum fruit tree), salicina (Japanese plum).
Pyrus	communis (pear fruit tree).
Quercus	all.
Robinia	psuedoacacia (black locust) not for any garden because of suckering habit.
Salix (willows)	all.
Schinas molle (peppercorn)	
Sorbus	all.
Tristania	conferta (brush box).
Ulmus (elm)	all.
Virgilia	all.
Waterhousea	florabunda (weeping lilly pilly).

Appendix 6

Wastewater reuse EPA guidelines

The Environmental Protection Authority (EPA) has identified acceptable and unacceptable wastewater reuse practices and the associated risks. It stresses, though, that the best way to reduce water use is to adopt household water conservation practices.

Every grey-water collection, treatment and reuse system must be approved by the EPA and be installed by a licensed plumber. It is the role of the manufacturer to obtain approval for the actual system, and the role of the householder to apply to the local council and water authorities for installation. There are penalties and civil liabilities for allowing grey water to pollute air, adjoining lands and waterways.

The EPA Victoria Information Bulletin 812 lays down the following implementation guidelines.

Grey water:

- must be used only during warm dry periods
- must not contaminate drinking water
- must have fail-safe diversion mechanisms in case of blockages
- must be diverted to the sewer during wet periods
- must not contain kitchen water
- can contain shower, bath, hand basin and laundry rinse water
- must not contain faecal matter, for example from nappy-washing
- should be applied to the garden by a subsurface system
- requires users to wash their hands after gardening
- must not be stored for more than 24 hours
- must not be allowed to attract rodents and insects
- should not be allowed to discharge outside the property
- should never be drunk by humans or animals

- should be used only on healthy plants
- should contain only low-phosphorus detergents
- should not be allowed to saturate the soil.

The EPA recommends reduced use of fertiliser on the garden when watering with grey water. The Information Bulletin also contains detailed water storage requirements and a checklist for such schemes, which includes management standards and setting up an irrigation management plan. Grey water reuse for toilet-flushing is the responsibility of local water authorities.

You must obtain a septic tank permit to install a reuse system.

Appendix 7

Water audit

The following information is made available courtesy of South West Water.

Spending 10–20 minutes completing this water audit will help you work out how much water you use in your household and garden. The audit has been tailored for individual rooms in your house. Write your audit sum in the box below each table. Combine your annual indoor water usage with your annual outdoor water usage, then complete the details at the end of the audit to determine the estimated total water usage figure per year for your household.

All figures used in the audit tables are approximate. They will vary depending on your type of appliances and the water pressure in your home.

All figures used in the tables have been derived from City West Water's *Water Audit Kit*, 1996.

Indoor usage

This 'indoor' water usage section targets each room in your house, allowing you to work out which room uses most water. It may highlight potential areas where your family can reduce water usage.

A section on running taps has also been included. This section audits the common tap used for rinsing dishes or clothes, for washing your face, or for drinking water.

In the kitchen

Washing dishes

Depth of water in sink	Number of times per week						For each extra wash add
	1	2	4	7	15	20	
Less than 1/2 full	400	800	1 600	2 800	6 000	8 000	400
Over 1/2 full	8000	1 6000	3 2000	5 6000	12 000	16 000	800

All figures are in L per year

Total water use for washing dishes = ☐ L per year

Dishwashers

Dishwasher cycle	Number of times per week						For each extra cycle add
	1	2	4	7	15	20	
Normal	2 000	4 000	8 000	14 000	30 000	40 000	2 000
Rinse/hold	500	1 000	2 000	3 500	7 500	10 000	500

All figures are in L per year

Total water use for dishwasher = ☐ L per year

The running tap (rinsing, drinking, bathroom)

Number of people in household	Average number of minutes a tap runs per day per person						For each extra minute add
	1	3	5	7	10	20	
One	750	2 000	3 500	5 000	7 500	15 000	750
Two	1 500	4 000	7 000	10 000	15 000	30 000	1500
Three	2 250	6 000	10 500	15 000	22 500	45 000	2250
Four	3 000	8 000	14 000	20 000	30 000	60 000	3000
For each extra person add	750	2 000	3 500	5 000	7 500	15 000	

All figures are in L per year

Note: An average figure of 5 minutes per person per day and a flow rate of 2 L/min has been assumed.

Total water use for running taps = ☐ L per year

Appendix 7 | 169

In the bathroom

Baths

Depth of water in bath	Number of baths per week						For each extra bath add
	1	2	4	7	15	20	
Less than 1/4 full	2 500	5 000	10 000	17 500	37 500	50 000	2 500
1/4–1/2 full	5 000	10 000	20 000	35 000	75 000	100 000	5 000
1/2–3/4 full	7 500	15 000	30 000	52 500	112 500	150 000	7 500
More than 3/4 full	10 000	20 000	40 000	70 000	150 000	200 000	10 000
All figures are in L per year							

Total water use for baths = ☐ L per year

Spa baths

Depth of water in spa bath	Number of baths per week						For each extra bath add
	1	2	4	7	15	20	
Less than 1/4 full	2 500	5 000	10 000	17 500	37 500	50 000	2 500
1/4–1/2 full	5 000	10 000	20 000	35 000	75 000	100 000	5 000
1/2–3/4 full	7 500	15 000	30 000	52 500	112 500	150 000	7 500
More than 3/4 full	10 000	20 000	40 000	70 000	150 000	200 000	10 000
All figures are in L per year							

Total water use for spa baths = ☐ L per year

Toilet

Type of toilet	Number of people in house						For each extra person add
	1	2	3	4	5	6	
Dual flush 6/3 L	10 000	18 000	25 000	30 000	34 000	39 000	5 000
Dual flush 11/5 L	18 000	32 000	44 000	53 000	61 000	70 000	8 000
Single flush 11 L	24 000	44 000	60 000	72 000	84 000	90 000	12 000
All figures are in L per year							

Total toilet water use = ☐ L per year

Showers

Ordinary shower rose (14 L/min)	Number of showers per week						For each extra shower add
	1	7	10	15	20	30	
3 minutes	2 100	14 700	21 000	31 500	42 000	63 000	2 100
5 minutes	3 500	24 500	35 000	52 500	70 000	105 000	3 500
10 minutes	7 000	49 000	70 000	105 000	140 000	210 000	7 000
20 minutes	14 000	98 000	140 000	210 000	280 000	420 000	14 000
For each extra minute add 700		4 900	7 000	10 500	14 000	21 000	
All figures are in L per year							

Total water use for ordinary shower rose = ☐ L per year

Water-efficient shower rose (8 L/min)	Number of showers per week						For each extra shower add
	1	7	10	15	20	30	
3 minutes	1 200	8 400	12 000	18 000	24 000	36 000	1 200
5 minutes	2 000	14 000	20 000	30 000	40 000	60 000	2 000
10 minutes	4 000	28 000	40 000	60 000	80 000	120 000	4 000
20 minutes	8 000	56 000	80 000	120 000	160 000	240 000	8 000
For each extra minute add 400		2 800	4 000	6 000	8 000	12 000	
All figures are in L per year							

Total water use for water-efficient shower rose = ☐ L per year

Shower flow rates vary from 6 to 30 L per minute. You can test your flow rate by placing a bucket under the shower for 10 seconds. Measure the amount of water in the bucket and multiply by 6 to convert to litres per minute. Use the following calculation to work out your actual usage in the shower:

☐ × ☐ × ☐ × ☐ = ☐
L/min   min per shower   showers per week   weeks per year   L/pa

Appendix 7 | 171

In the laundry

Washing machine

Type of washing machine	Number of washes per week						For each extra wash add
	1	2	4	7	15	20	
Twin tub	2 000	4 000	8 000	14 000	30 000	40 000	2 000
Wringer	3 000	6 000	12 000	21 000	45 000	60 000	3 000
Front loader	6 000	12 000	24 000	42 000	90 000	120 000	6 000
Small automatic	6 000	12 000	24 000	42 000	9 0000	120 000	6 000
Medium automatic	7 000	14 000	28 000	49 000	105 000	140 000	7 000
Large automatic	10 000	20 000	40 000	70 000	150 000	200 000	10 000
All figures are in L per year							

If you use other settings on all washes: subtract 20% for low water level; subtract 10% for medium water level; subtract 25% for suds saver.

Total water use for washing machine = ☐ L per year

Laundry sink

Depth of water in laundry sink	Number of sinks per week						For each extra sink add
	1	2	4	7	15	20	
Less than 1/2 full	1 000	2 000	4 000	7 000	15 000	20 000	1 000
Over 1/2 full	2 000	4 000	8 000	14 000	30 000	40 000	2 000
All figures are in L per year							

Total water use for laundry sink = ☐ L per year

Calculating your indoor water usage

Add up all the indoor water usage audit results.
Your total indoor water usage is ☐ L per year.

Outdoor usage

This aspect of your water audit is designed to help you calculate how much water you use annually in the garden, to wash the car or fill the pool.

The garden section has been divided into seasonal categories as your garden needs more water in summer than in winter and sometimes needs extra watering sessions, depending on the season.

Watering garden and lawns

Using hand-held hose

Frequency of watering: Summer	Number of minutes per session						For each extra 10 minutes add
	10	30	60	90	120	180	
Twice a day	40 000	120 000	240 000	360 000	480 000	720 000	40 000
Once a day	20 000	60 000	120 000	180 000	240 000	360 000	20 000
Once every 2 days	10 000	30 000	60 000	90 000	120 000	180 000	10 000
Once every 3 days	6 000	18 000	36 000	54 000	72 000	108 000	6 000
Once a week	2 800	8 400	16 800	25 200	33 600	50 400	2 800
Once a fortnight	1 400	4 200	8 400	12 600	12 800	25 200	1 400
All figures are for summer period only							

Total summer watering by hose = ☐ L per year

Frequency of watering: Autumn and Spring	Number of minutes per session						For each extra 10 minutes add
	10	30	60	90	120	180	
Once a day	30 000	90 000	180 000	270 000	360 000	540 000	30 000
Once every 2 days	15 000	45 000	90 000	135 000	180 000	270 000	15 000
Once every 3 days	10 000	30 000	60 000	90 000	120 000	180 000	10 000
Once a week	4 800	14 400	28 800	43 200	57 600	86 400	4 800
Once a fortnight	2 400	7 200	14 400	21 600	28 800	43 200	2 400
Once a month	1 200	3 600	7 200	10 800	14 400	21 600	1 200
All figures are for autumn/spring period only							

Total autumn/spring watering by hose = ☐ L per year

Frequency of watering: Winter	Number of minutes per session						For each extra 10 minutes add
	10	30	60	90	120	180	
Once a day	20 000	60 000	120 000	180 000	240 000	360 000	20 000
Once every 2 days	10 000	30 000	60 000	90 000	120 000	180 000	10 000
Once every 3 days	6 000	18 000	36 000	54 000	72 000	108 000	6 000
Once a week	2 800	8 400	16 800	25 200	33 600	50 400	2 800
Once a fortnight	1 400	4 200	8 400	12 600	16 800	25 200	1 400
Once a month	700	2 100	4 200	6 300	8 400	12 600	700
All figures are for winter period only							

Total winter watering by hose = ☐ L per year

Using bucket or watering-can

Frequency of watering: Summer	Number of buckets per session						For each extra bucket add
	1	5	10	15	20	30	
Twice a day	1 800	9 000	18 000	27 000	36 000	54 000	1 800
Once a day	900	4 500	9 000	13 500	18 000	27 000	900
Once every 2 days	450	2 300	4 500	6 800	9 000	13 500	450
Once every 3 days	250	1 300	2 500	3 800	5 000	7 500	250
Once a week	150	750	1 500	2 300	3 000	4 500	150
Once a fortnight	75	350	750	1 100	1 500	2 300	75
All figures are for summer period only							

Total summer watering by bucket or watering-can = ☐ L per year

Frequency of watering: Autumn and Spring	Number of buckets per session						For each extra bucket add
	1	5	10	15	20	30	
Once a day	1 350	6 800	13 500	20 300	27 000	40 500	1 350
Once every 2 days	680	3 400	6 800	10 200	13 600	20 400	680
Once every 3 days	450	2 300	4 500	6 800	9 000	13 500	450
Once a week	220	1 100	2 200	3 300	4 400	6 600	220
Once a fortnight	110	600	1 100	1 700	2 200	3 300	110
Once a month	50	250	500	750	1 000	1 500	50
All figures are for autumn/spring period only							

Total autumn/spring watering by bucket or watering-can = ☐ L per year

Waterwise House & Garden

Frequency of watering: Winter	Number of buckets per session						For each extra bucket add
	1	5	10	15	20	30	
Once a day	900	4 500	9 000	13 500	18 000	27 000	900
Once every 2 days	450	2 300	4 500	6 800	9 000	13 500	450
Once every 3 days	250	1 300	2 500	3 800	5 000	7 500	250
Once a week	150	750	1 500	2 300	3 000	4 500	150
Once a fortnight	75	350	750	1 100	1 500	2 300	75
Once a month	30	150	300	450	600	900	30
All figures are for winter period only							

Total winter watering by bucket or watering-can = [   ] L per year

Using hose connected to garden sprinkler

Soaker hoses, rotating jets, rose and wave-type sprinklers, and micro-sprays are included.

Frequency of watering: Summer	Number of minutes per session						For each extra 10 minutes add
	10	30	60	90	120	180	
Twice a day	30 000	90 000	180 000	270 000	360 000	540 000	30 000
Once a day	15 000	45 000	90 000	135 000	180 000	270 000	15 000
Once every 2 days	7 500	22 500	45 000	67 500	90 000	135 000	7 500
Once every 3 days	4 500	13 500	27 000	40 500	54 000	81 000	4 500
Once a week	2 000	6 000	13 000	18 000	25 000	36 000	2 000
Once a fortnight	1 000	3 000	6 500	9 000	12 500	18 000	1 000
All figures are for summer period only							

Total summer watering by sprinkler = [   ] L per year

Frequency of watering: Autumn and Spring	Number of minutes per session						For each extra 10 minutes add
	10	30	60	90	120	180	
Once a day	22 500	67 500	135 000	202 500	270 000	405 000	22 500
Once every 2 days	11 000	33 000	66 000	99 000	133 000	198 000	11 000
Once every 3 days	7 500	22 500	45 000	67 500	90 000	135 000	7 500
Once a week	3 500	10 500	21 000	31 500	42 000	63 000	3 500
Once a fortnight	1 800	5 400	10 800	16 200	21 600	32 400	1 800
Once a month	900	2 700	5 400	8 100	10 800	16 200	900
All figures are for autumn/spring period only							

Total autumn/spring watering by sprinkler = [   ] L per year

Appendix 7 | 175

Frequency of watering: Winter	Number of minutes per session						For each extra 10 minutes add
	10	30	60	90	120	180	
Once a day	15 000	45 000	90 000	135 000	180 000	270 000	15 000
Once every 2 days	7 500	22 500	45 000	67 500	90 000	135 000	7 500
Once every 3 days	4 500	13 500	27 000	40 500	54 000	81 000	4 500
Once a week	2 000	6 000	13 000	18 000	25 000	36 000	2 000
Once a fortnight	1 000	3 000	6 500	9 000	12 500	18 000	1 000
Once a month	500	1 500	3 000	4 500	6 000	9 000	500
All figures are for winter period only							

Total winter watering by sprinkler = ☐ L per year

Using fixed sprinklers and pop-up systems

Frequency of watering: Summer	Number of minutes per session						For each extra 10 minutes add
	10	30	60	90	120	180	
Twice a day	60 000	180 000	360 000	540 000	720 000	1 080 000	60 000
Once a day	30 000	90 000	180 000	270 000	360 000	540 000	30 000
Once every 2 days	15 000	45 000	90 000	135 000	180 000	270 000	15 000
Once every 3 days	9 000	27 000	54 000	81 000	108 000	162 000	9 000
Once a week	4 000	12 000	24 000	36 000	48 000	72 000	4000
Once a fortnight	2 000	6 000	12 000	18 000	24 000	36 000	2 000
All figures are for summer period only							

Total summer watering by fixed system = ☐ L per year

Frequency of watering: Autumn and Spring	Number of minutes per session						For each extra 10 minutes add
	10	30	60	90	120	180	
Once a day	45 000	135 000	270 000	405 000	540 000	810 000	45 000
Once every 2 days	22 500	67 500	135 000	202 500	270 000	405 000	22 500
Once every 3 days	15 000	45 000	90 000	135 000	180 000	270 000	15 000
Once a week	7 000	21 000	42 000	63 000	84 000	126 000	7 000
Once a fortnight	3 500	10 500	21 000	31 500	42 000	63 000	3 500
Once a month	1 800	5 400	10 800	16 200	21 600	32 400	1 800
All figures are for autumn/spring period only							

Total autumn/spring watering by fixed system = ☐ L per year

Waterwise House & Garden

Frequency of watering: Winter	Number of minutes per session						For each extra 10 minutes add
	10	30	60	90	120	180	
Once a day	30 000	90 000	180 000	270 000	360 000	540 000	30 000
Once every 2 days	15 000	45 000	90 000	135 000	180 000	270 000	15 000
Once every 3 days	9 000	27 000	54 000	81 000	108 000	162 000	9 000
Once a week	4 000	12 000	24 000	36 000	48 000	72 000	4 000
Once a fortnight	2 000	6 000	12 000	18 000	24 000	36 000	2 000
Once a month	1 000	3 000	6 000	9 000	12 000	18 000	1 000
All figures are for winter period only							

Total winter watering by fixed system = ☐ L per year

Using drip system

Frequency of watering: Summer	Number of minutes per session						For each extra 30 minutes add
	60	120	180	240	300	360	
Once a day	30 000	60 000	90 000	120 000	150 000	180 000	15 000
Once every 2 days	15 000	30 000	45 000	60 000	75 000	90 000	7 500
Once every 3 days	9 000	18 000	27 000	36 000	45 000	54 000	4 500
Once a week	4 000	8 000	12 000	16 000	20 000	24 000	2 000
Once a fortnight	2 000	4 000	6 000	8 000	10 000	12 000	1 000
Once a month	1 000	2 000	3 000	4 000	5 000	6 000	500
All figures are for summer period only							

Total summer watering by drip system = ☐ L per year

Frequency of watering: Autumn and Spring	Number of minutes per session						For each extra 30 minutes add
	60	120	180	240	300	360	
Once a day	45 000	90 000	135 000	180 000	225 000	270 000	22 500
Once every 2 days	22 500	45 000	67 500	90 000	112 500	135 000	11 250
Once every 3 days	15 000	30 000	45 000	60 000	75 000	90 000	7 500
Once a week	7 000	14 000	21 000	28 000	35 000	42 000	3 500
Once a fortnight	3 500	7 000	10 500	14 000	17 500	21 000	1 750
Once a month	1 750	3 500	5 250	7 000	8 750	10 500	875
All figures are for autumn/spring period only							

Total autumn/spring watering by drip system = ☐ L per year

Appendix 7 | 177

Frequency of watering: Winter	Number of minutes per session						For each extra 30 minutes add
	60	120	180	240	300	360	
Once a day	30 000	60 000	90 000	120 000	150 000	180 000	15 000
Once every 2 days	15 000	30 000	45 000	60 000	75 000	90 000	7 500
Once every 3 days	9 000	18 000	27 000	36 000	45 000	54 000	4 500
Once a week	4 000	8 000	12 000	16 000	20 000	24 000	2 000
Once a fortnight	2 000	4 000	6 000	8 000	10 000	12 000	1 000
Once a month	1 000	2 000	3 000	4 000	5 000	6 000	500
All figures are for winter period only							

Total winter watering by drip system = ☐ L per year

Washing paths and driveways

Using hand-held hose

Frequency of watering	Number of minutes per session						For each extra 30 minutes add
	10	30	60	90	120	180	
Twice a day	40 000	120 000	240 000	360 000	480 000	720 000	40 000
Once a day	20 000	60 000	120 000	180 000	240 000	360 000	20 000
Once every 2 days	10 000	30 000	60 000	90 000	120 000	180 000	10 000
Once every 3 days	6 000	18 000	36 000	54 000	72 000	108 000	6 000
Once a week	2 800	8 400	16 800	25 200	33 600	50 400	2 800
Once a fortnight	1 400	4 200	8 400	12 600	12 800	25 200	1 400
All figures are in L per year							

Total water used on washing paths and driveways by hose = ☐ L per year

Swimming pools

Filling up

Size of pool	Number of times filled per year						For each extra fill add
	1	2	3	4	5	10	
Wading pool	500	1 000	1 500	2 000	2 500	5 000	500
Small pool	20 000	40 000	60 000	80 000	-	-	20 000
Medium pool	50 000	100 000	150 000	-	-	-	50 000
Large pool	110 000	220 000	-	-	-	-	110 000
All figures are in L per year							

Total water used for filling pools = ☐ L per year

Topping-up

	Number of hours topping-up per year						For each extra fill add
	1	2	5	10	20	30	
Any sized pool	700	1 400	3 500	7 000	14 000	21 000	700
All figures are in L per year							

Total water used for topping-up pools = ☐ L per year

Washing car

Using bucket

Number of buckets used per wash	Number of times car is washed per year						For each extra 10 washes add
	10	20	50	100	200	300	
One bucket	70	140	350	700	1 400	2 100	70
Two buckets	140	280	700	1 400	2 800	4 200	140
Three buckets	210	420	1 050	2 100	4 200	6 300	210
Four buckets	280	560	1 400	2 800	5 600	8 400	280
For each extra bucket add	70	140	350	700	1 400	2 100	
All figures are in L per year							

Total water used for washing car by bucket = ☐ L per year

Using hose

Number of minutes per wash	Number of times car is washed per year						For each extra 10 washes add
	10	20	50	100	200	300	
5 minutes	500	1 000	2 500	5 000	10 000	15 000	500
10 minutes	1 000	2 000	5 000	10 000	20 000	30 000	1 000
15 minutes	1 500	3 000	7 500	15 000	30 000	45 000	1 500
30 minutes	3 000	6 000	15 000	30 000	60 000	90 000	3 000
60 minutes	6 000	12 000	30 000	60 000	120 000	180 000	6 000
For each extra 5 minutes add	500	1 000	2 500	5 000	10 000	15 000	
All figures are in L per year							

Total water used for washing car by hose = ☐ L per year

Calculating your outdoor water usage
Add up all the outdoor water usage audit results. Your total outdoor water usage is ☐ L per year.

Calculating your household's estimated total annual water usage
Your total indoor water usage = ☐ L per year
Your total outdoor water usage = ☐ L per year
Your household's estimated total annual water usage = ☐ L per year

Now that you are aware of how your household uses water, you will be able to take a few measures to reduce that water usage.

Carry out another water audit in approximately 12 months and compare the results with your current result. They may be quite different, due to a number of factors.

- Your answers are either under- or over-estimated.
- You are using different or new appliances.
- Your water pressure varies.
- There are more or few people living in your household.

Try to achieve an annual water usage less than your original one.

Remember to check your water meter regularly for early detection of leaking taps, cisterns or appliances.

Appendix 8

Publications

Be careful with books from overseas. When they say 'dry' they do not necessarily mean the 'dry' we experience in Australia, which is generally harsh and extended.

Australian natives

Australian Plant Study Group (1997) *Grow what where*. Lothian, Melbourne. An excellent book for garden design.

Elliot, W.R. & Jones, D.L. (1980–94) *Encyclopaedia of Australian plants suitable for cultivation. Vols 1-8*. Lothian, Melbourne. This is the most comprehensive, well-illustrated, detailed reference of Australian plants. This series (not yet complete) represents a life of dedicated work and might be too expensive for all but the most enthusiastic plant lover. However, your local library should have the set.

Stones, E. (1971) *Australian garden design*. Macmillan, Melbourne.

Wrigley, J.W. (1996) *Australian native plants: propagation, cultivation and use in landscaping*, 4th edn. Reed Books, Sydney.

There are many localised publications by native plant groups and you are almost certain to find one for your area. Start your enquiries at your local botanic garden.

Waterwise garden books

Clayton, S. (1994) *The reverse garbage mulch book*. Hyland House, Melbourne. A good practical book by a dedicated gardener.

Taylor, J. *The Dry Garden*. Lothian, Melbourne. Written for the international market, this book has detailed descriptions of many plants and photos of some.

van Dok, W. (2000) *The water-efficient garden: includes detailed information on greywater irrigation*. Wendy van Dok, Glen Waverley, Melbourne. Its focus is on water use and reuse in gardens of southern Australia, but it is also useful elsewhere in Australia.

Walsh, K. (1995) *Water-saving gardening: water-wise plants and practices in Australia.* Reed Books, Sydney. A good-quality book covering design, mulching, soil improvement and plant selection, mainly for southern Australia.

Herbs

Fletcher, K. (1996) *The Penguin modern Australasian herbal*, rev. edn. Penguin, Melbourne.

Lust, J. *The herb book.* Bantam.

Patterson, D. (ed.) *Culpeppers color herbal.* Well-illustrated.

Stuart, M. (ed.) (1979) *The encyclopedia of herbs and herbalism.* Orbis Books, London. Excellent, well-illustrated all-purpose book.

These are just a few of the books available on herbs; your local library will have more. From a waterwise perspective, look for herbs of Mediterranean and Middle Eastern origin or from deserts in other parts of the world.

Earthworms

Clayton, S. (1994) *The reverse garbage mulch.* Hyland House, Melbourne.

Roads, M.J. (1989) *The natural magic of mulch: organic gardening Australian style.* Greenhouse Publications, Melbourne.

Taylor, D. & Taylor, Y. (1993) *The compost book.* Reed Books, Sydney.

Windust, A. (1997) *Worms garden for you.* Allscape, Mandurang. For the home gardener.

Windust, A. (1997) *Worms downunder downunder: for farm, garden, schools, profit and recycling.* Allscape, Mandurang. A general introduction to earthworms.

Windust, A. (1997) *Worm farming made simple.* Allscape, Mandurang. For amateurs to professionals.

Windust, A. (2000) *Green home recycling: 16 ways to compost or worm farm, landcare for every household.* Allscape, Mandurang. A book for every household.

Allscape can be contacted by phone on (03) 5439 5099.

Water reuse

Jeppesen, B. (1993) *Domestic greywater reuse: preliminary evaluation.* Research Report No. 60, Urban Water Research Association of Australia. Melbourne Water Corporation, Melbourne.

Jeppesen, B. & Solley D. (1994) *Domestic greywater reuse: overseas practice and its applicability to Australia.* Research Report No. 73, Urban Water Research Association of Australia. Melbourne Water Corporation, Melbourne.

Jeppesen, B. (1996) *Model guidelines for domestic greywater reuse for Australia.* Research Report No. 107, Urban Water Research Association of Australia. Melbourne Water Corporation, Melbourne.

Only the most enthusiastic student would need all three volumes. Interested householders would probably need only the *Model guidelines for domestic greywater reuse for Australia.*

General

Villiers, M. de. (1999) *Water wars: is the world's water running out.* Weidenfeld & Nicolson, London.

About the author

Allan Windust

Allan Windust has been a gardener for nearly 50 years and a garden designer in urban and country areas for over 30 years. He has experienced several droughts. The most recent drought has made a great impression upon him, especially during his travels through coastal and inland Queensland and New South Wales.

Allan has had an extensive professional career covering a variety of roles:

- a professional surveyor, reaching the level of District Surveyor
- an environmental planner with the Land Conservation Council and Department of Natural Resources and Environment
- manager of a Landcare Information Centre
- environment consultant to various bodies such as the Australian Government Solicitor.

Allan is also the author of several books on aspects of the environment:

- *Buying your bush block*
- *Worms downunder downunder*
- *Worm farming made simple*
- *Worms garden for you*
- *Green home recycling: 16 ways to compost or worm farm*
- *Permaculture for beginners.*

New titles that Allan is working on are:

- *Eat your organic garden*
- *Small farm design.*

He also runs weekend courses on:

- dry stone walling
- organic gardening
- stone construction
- worm farming

Enquiries

For more details please send a stamped self-addressed envelope to:

Allan Windust
Allscape
38 Ronald Drive
Mandurang Vic. 3551
Phone: (03) 5439 5099
Fax: (03) 5439 3400
Email: allscape@bigpond.com.au

Author's note

You are invited to contribute your ideas, photos and suggestions to this and any other of my books.

Index

acacias 106
Australian Plant Society 111, 112
Australian plants for dry conditions 106–107, 135–144
automatic watering systems 29
bagasse 68
banksias 107
bathroom water use 169–170
baths 20, 40, 169
biofiltration 123
bird attraction 9
black water 38, 41, 119–120
bossiaeas 107
bracken 68–69
bucket watering 21, 50, 173–174
callistemons 107
callitris 107
car washing 178
carbon sink, suburban 8–9
carpet 69
casuarinas 106
chipwood 69
chorizemas 107
climate, Australian 11–13
   local 15–16, 87–91
climate change 1–2
clover hay 72–73
coffee grounds 70
compost 70
composting toilet 38

correas 107
cow manure 70
dams 37
desalinisation 124
dillwynias 107
dishwashers 20, 41, 168
drainage 91, 101–102
drains, damage by plants 163–164
drinking water, from water tank 32
drip watering systems 176–177
dripper hoses 26
drippers, individual 27
driveways 177
drought, 11, 13, 113–117
   gardening during 113–117
   planning for 114–115
   watering routine during 114
drought-tolerant plants 105–108, 135–159
dry seasons see drought
dry zones, in garden 95
earthworms 55–56
elevated water tank 34
eucalyptus mulch 70–71
evapotranspiration 15–16, 17
exotic plants, drought-tolerant 107–108, 145–159
feathers 71
filters 27
fire buffer zones 97
fire planning 93, 161

fire-retardant garden design   97–99
fire-retardant plants   161–162
flowers   129
foliage watering   51
frost   50, 61, 126
fruit   129
garden, climate   87–91
    design   95–100, 102–104
    planning   83–104
    structures   91
    water recycling   101–102
    water usage   172–177
    watering methods   21–30, 49–52
    zones   95–99
gardening, waterwise rules   116–117
grass clippings *see* lawn clippings
grass hay   73
grasses   107
green manure   57
grevilleas   107
grey water,   31, 38–42, 120–121
    attitudes to use   120–121
    reuse   48–49
    reuse guidelines   41–42, 48, 165–166
    reuse systems   42–47
    use on gardens   39, 48
hair   72
hakeas   107
hand basins   20, 41
hay   72–73
headenbergias   107
heat stress, and watering   50–51
hessian   73
hose, watering by   21–22, 172–173, 177
hoses, dripper   26
    porous   25–26
household wastewater recycling   37–49
household reticulation options   19–20
indoor water usage   167–171
industrial wastes   67
kennedias   107
kitchen garden   99–100
kitchen sinks   20, 41, 168
kitchen, water use   168
lake weed   73–74
laundry sink   20, 40, 171
laundry, water use   171
lawn clippings   66, 71–72
lawns   51, 95–97, 114, 161

leaf litter   74
leaves   128, 132
lucerne hay   72
lupin hay   73
mains water   17
manures   70, 74, 76, 77, 80
melaleucas   107
micro-climate zones   87–89
misters   24–25
moisture meters   28–29
moisture testing   49
mulch,   53–81
    applying   58–62
    collecting   67
    depth   53–54
    gravel or stone   63
    as habitat   55–56
    as habitat for pests   67–68
    as insulation   54–55
    living   63–64
    non-organic   62–63
    role of   53
    as soil protector   56
    types of   62–63, 65–81
    as weed suppressor   57
mushroom compost   75
myoporum   107
noise control   92
nurseries   110
on-ground water tank   34–35
outdoor water usage   172–179
paper   75
paths   177
pea straw   76
peanut shells   75–76
peat moss   76
pig manure   76
pine needles   76–77
plant anatomy   127–130
plant cells   129–130
plant selection   109–112, 114
plants,   125–134
    damage to drains   163–164
    drought-tolerant   105–108, 135–144
    fire-retardant   161–162
    self-mulching   64–65
    survival strategies   130–134
    and water stress   125–126
plastic   77

ponds   37
porous hoses   25–26
pot plants   29–30, 116
poultry litter   77
privacy   92
prunings, as mulch   65–66
rainfall   1–2, 16
rice hulls   78
root watering   51
roots   127, 130–131
salinity   9–10
sawdust   78–79
scoria   79
seaweed   79–80
seepage losses, from garden   17
seepage water, recycling of   18
self-mulching plants   64–65
septic tank effluent   41
shade   89, 114
sheep manure   80
showers   19–20, 40, 170
site assessment plan   85–95
soil moisture   126–127
soil protection, by mulch   56
sprinklers   22–24, 174–176
stable manure   80
stems   128, 131
stormwater   124
suburban water usage   18
swimming pools   35, 177–178
tanbark   81
tank water, for toilet cisterns   36
taps   168
temperature   90
timers, water   18, 28
toilets,   19, 36, 169
    composting   38
underground water tank   35–36
underlay   69
washing dishes   168

washing machine   20, 40, 171
wastewater,
    future use   119–124
    recycling   37–49
    reuse guidelines   165–166
    reuse systems   122–124
    use and public health   121
water audit   167–179
water budget   52
water recycling, garden water   101–102
water stress   125–126
water supply, during drought   113
water tanks,   16–17, 30–37
    capacity   31–32
    construction   33
    and drinking water   32
    inspection   33–34
    installation   32–33
    location   34–35
    maintenance   33–34
    regulations   30–31
water use,
    bathroom   169
    indoor   167–171
    kitchen   168
    laundry   171
    outdoor   172–179
    quantity   17
    suburban   18
watering, methods   21–30, 49–52
    routine during drought   114
    times of   52
    water use   172–177
weed suppression   57
weeds, as green manure   57
wet zones, in garden   95
wind   90–91, 92
wood ash   81
wood shavings   81